Engraving Accuracy in Early Modern England

Engraving Accuracy in Early Modern England

Visual Communication and the Royal Society

Meghan C. Doherty

Routledge
Taylor & Francis Group
LONDON AND NEW YORK

First published in 2022 by Amsterdam University Press Ltd.

Published 2025 by Routledge
4 Park Square, Milton Park, Abingdon, Oxon OX14 4RN
605 Third Avenue, New York, NY 10158

Routledge is an imprint of the Taylor & Francis Group, an informa business

ISBN: 9789463721066 (hbk)
ISBN: 9781041178781 (pbk)
ISBN: 9781003694816 (ebk)
NUR 685

Cover illustration: Anonymous, *How to draw a sphere and a head*, facing page 120, Engraving, from John Evelyn, *Sculptura* (London: Printed by J. C. for G. Beedle & T. Collins, 1662). Bodleian Library, 8vo G.119 Linc.

Cover design: Coördesign, Leiden

DOI 10.5117/9789463721066

For Product Safety Concerns and Information please contact our EU representative:
GPSR@taylorandfrancis.com
Taylor & Francis Verlag GmbH, Kaufingerstraße 24, 80331 München, Germany

Table of Contents

Acknowledgements 7

Introduction 11
"Claiming the Resemblance of Life" 11
"The Best in the World in this kind" 15
"An accurate impression is in far higher esteem" 26
"Each Judgement of his Eye" 33
"Examining it according to my usual manner" 36

1. **"Innocent Witch-craft of Lights": Developing Visual Judgment through Printed Books** 49
The Magic of Projection 57
"Draw and Engrave their Schemes with Delight and Assurance" 68
"A fit subject for our kingdomes knowledge and practice" 77
Conclusion 89

2. **"A New Visible World": Developing a Visual Vocabulary for the Microscopic** 97
The Visual Culture of Early Microscopy 106
Developing a "sincere Hand and a faithful Eye" 113
Making "a Plain Representation" 117
Engraving "the True Form" 122

3. **"Nearly Resembling the Live Birds": Collecting and Collating for the Reformation of Natural History** 137
Resembling the Text: The Dodo (Raphus cucullatus) 145
Resembling the Printed Record: The Golden Eagle (Aquila chrysaetos) 149
Resembling the Living: The Smew (Mergus albellus) 157
Resembling the Dead: The Great Grey Gull 162
Conclusion 166

4. **"These Rude Collections": Accumulating Observations and Experiments** 177
"The present figure of *Saturn*" 182
"With so much care and exactness" 199
Conclusion 209

Conclusion 215

Bibliography 221

Index 239

Acknowledgements

Throughout this whole process one voice has been constant, both out loud and in my head. Every time our paths crossed over the years, Sachiko Kusukawa would always ask when she would be able to put my book on her shelf. Her faith in me and this project has helped me to keep going through all the trials and tribulations that have come my way. I am grateful that her voice was joined by others, especially Beth Yale, Anna Marie Roos, and Daniela Bleichmar. I do not think this project would have ever been completed if I had not had the voices of these strong women ringing in my ears.

This project began in Madison, where Jill H. Casid was my adamant and steadfast supporter. Robin Rider and Florence Hsia joined their voices with Jill's and encouraged and challenged my thinking throughout the research and writing of this book. I am also grateful for the early conversations and feedback provided by Lynn Nyhart, Tom Broman, and Henry Turner. Even though my research interests strayed away from early America, everyone at the Chipstone Foundation remained committed to seeing me succeed. I am particularly grateful to Jon Prown, Glenn Adamson, and Amy Meyers. Finally, although he could not understand why a person would want to learn to engrave in the twenty-first century, Jack Damer welcomed me into his introductory printmaking class, so I could gain a physical understanding of what Faithorne was writing about in *The Art of Graveing.*

Many libraries and even more librarians and curators opened their doors, vaults, and study rooms to me. At the University of Wisconsin–Madison, Robin Rider has been tireless in her support of my research and this project. In addition, at the then Elvehjem, now Chazen Museum, Drew Stevens was an early supporter of my research on William Faithorne. At multiple stages in this process my research has been supported by the Huntington Library, Art Museum, and Botanical Gardens. Both times I was there as a fellow, Dan Lewis and Stephen Tabor were endlessly patient and supportive with my many questions. Four months at the Smithsonian Libraries was essential to the research for this project. At the Dibner Library of the History of Science and Technology, Lila Vekerdy and Kirsten van der Veen helped me balance research and writing. At the Cullman Library of Natural History, Daria Wingreen-Mason and Leslie K. Overstreet pulled every book with an illustration of a bird published before 1676, which was essential for my research on the sources of the images in Willughby and Ray's *Ornithology.* My research on the *Philosophical Transactions* would not have been possible without the support of the staff at the Royal Society, in particular, Felicity

Henderson, Rupert Baker, and Keith Moore. A chance encounter with Giles Mandelbrot at the reference desk in the Rare Books Reading Room at the British Library led to many great conversations and the opportunity to share my work on *Albert Dürer Revived* and its connections to the London book trade. My research on that book, its many editions, and its sources was supported by a research fellowship at the Yale Center for British Art, which also allowed me time at the Beinecke Library where I met Kathryn James who has continued to support and encourage my research. Kim Sloan at the British Museum supported this project in the Print Room and with lots of conversations. Additional research took me to the National Art Library at the Victoria and Albert Museum, Stationer's Company Archives, the Folger Shakespeare Library, the Library of Congress, the Royal College of Physicians, the University of Nottingham, and Brown University's John Hay Library. Without these many patient librarians and curators, this project would not have been possible.

At the urging of Donna Sy (and without a clear understanding of what exactly critical bibliography was), I applied for the Andrew W. Mellon Fellowship of Scholars in Critical Bibliography at Rare Book School. Little did I know how much being a part of this fellowship would affect my scholarship. The first course I took at RBS was with Roger Gaskell and Caroline Duroselle-Melish. This course not only provided me with a wealth of information on the illustrated scientific books, but it also allowed me to meet the two wonderful instructors and the eleven other students. The conversations that came out of the class, particularly with Roger, Caroline, Dahlia Porter, and Courtney Roby, have shaped this project in innumerable ways. I am so grateful for their continued support. Conversations with Michael Suarez, Donna Sy, Barbara Heritage, and other RBS staffed have dramatically altered how I think about the book as a physical object as well as a container of knowledge. The introduction to this book benefitted greatly from feedback provided by fellow Fellows.

Many of the ideas in this book were first tested out in research projects over the years. I am grateful to the many scholars who shared their thoughts on these ideas along the way, in particular, Jim Bennet, Tim Birkhead, Maria Conforti, Mordechai Feingold, Sietske Fransen, Craig Ashley Hanson, Klaus Hentschel, Matthew Hunter, Michael Hunter, Jeanne Peiffer, and Katie Reinhart. These conversations, and many others, have helped shaped this project as they challenged my ideas about the materials presented.

Without the support of Erika Gaffney, at Amsterdam University Press, and Stefano Gulizia, Scientiae Series editor, this research project would never have become the book you are reading. I am grateful for their encouragement throughout the review process. This project benefitted greatly from two anonymous readers' feedback. Their comments helped me to strengthen the argument throughout the book. While many have read and commented on this along the way, any errors that remain are, of course, my own.

I thank the following institutions for their financial support of this project: University of Wisconsin–Madison, the Chipstone Foundation, the Yale Center for British Art, the Huntington Library, Art Museum, and Botanical Gardens, the Smithsonian Dibner Library of the History of Science and Technology, Andrew W. Mellon Fellowship of Scholars in Critical Bibliography, Rare Book School, Byrne-Bussey Marconi Fellowship at the Bodleian Libraries, Oxford University, Lisa Jardine History of Science Grant, The Royal Society, and Berea College.

Support for the image permission fees was provided by Berea College. I thank the following institutions for granting me the rights to reproduce images in their collections: the Ashmolean Museum, the Bodleian Library, the British Library, the British Museum, the Huntington Library, the Natural History Museum, London, the Royal Society, the Tate Britain, and the Vatican Library. Furthermore, I am grateful to the follow institutions that provided images at no cost: the Biodiversity Heritage Library, Brown University, the Chazen Museum, e-rara, Ebling Library at the University of Wisconsin–Madison, the National Gallery of Art, Washington, D.C., the National Portrait Gallery, London, Special Collections at the University of Wisconsin–Madison, and the Yale Center for British Art. I am also grateful to the editors of *Notes and Records of the Royal Society* for allowing me to include a heavily revised version of an essay first published there.

While this project was started in Madison, it came to a conclusion in Berea. Over the years, I have been endlessly grateful to have Dan Feinberg as an interlocutor in the drafting and editing of this project. Through the College's Labor Program, I have had the help of a number of students. In particular, I am grateful to: Haley Boothe, Elizabeth Ronilo, and Megan McEahern.

This project would never have begun without the support of my parents, Katie and Craig. Their enduring encouragement and willingness to read and copy edit endless drafts have sustained me throughout this process. I would never have finished this project without the love and support of Jason and Ellis. Jason has helped me keep going even when I wanted to give up completely and convinced me that people other than family really did care about what I had to say. Although this started before she was born, Ellis quickly began making her own books alongside me encouraging me to finish my own.

Introduction

Accuracy is not a characteristic that is innate in an image. It does not simply reside there, perfect and unchanging. Instead it is made. Made by hand. Made by a person. In the case of early modern scientific illustrations, it was often made with a burin. The engraver used a burin to carve lines into a prepared copper plate. These lines were then inked and printed onto dampened paper. As the paper moved through the press, the rollers pressed the damp paper into the carved lines. The ink was soaked up into the paper and an intaglio print was made. Whether accuracy was achieved depended on many variables: who made it, how he made it, where it appeared, how the surrounding text described it, what other images the viewer had seen. These variables could be controlled and needed to be to produce the effect of accuracy.

Early modern scholars and artisans worked together and separately to preclude the introduction of errors at every step. This is not to say that all the images then looked the same. On the contrary, accuracy implied careful methods of production, not visual consistency across instances of representation. Accuracy was constructed through a series of decisions by the author/artist and perceived by the reader through a series of judgments based on a reader's experience of the world, an image's relationship to other images, and the text accompanying it. The perception of an image's accuracy was the product of participation in a shared visual culture. Visual training was a precursor to the mutual acceptance of the accuracy of images. This training operated across a range of disciplines and through an unrecognized, but significant genealogy of image makers, consumers, and contexts. To understand how accuracy developed as a desired trait in an image, this book examines intaglio printed images produced by and for the Fellows of the Royal Society of London from 1660 to 1680.

"Claiming the Resemblance of Life"

Founded in 1660 for the promotion of experimental learning, the Royal Society brought together like-minded individuals intent on gaining a better

Doherty, M.C., *Engraving Accuracy in Early Modern England: Visual Communication at the Early Royal Society*. Taylor & Francis Group, 2022.
DOI 10.5117/9789463721066_INTRO

understanding of nature through the collection of observations and the production of experiments.[1] Fellows published the results of their experiments and observations both in single-author works and in the *Philosophical Transactions*, which in this period were the unofficial corporate record of their activities. These books and journal articles included intaglio printed images to aid in the transmission of knowledge to a broad audience. Engravers were hired to carve a visual record of these experiments and observations into copper plates to create printed images that could circulate widely.

Engraving Accuracy investigates artists' manuals alongside scientific treatises to probe how accuracy was produced and, subsequently, how images could be useful for early modern natural historians and natural philosophers. Manuals for learning to draw and engrave are analyzed alongside a history of engraving to set the groundwork for understanding how images were made and, more importantly, how careful actions were stressed as necessary precursors to successful image-making, which builds on the increased scholarly attention to printmaking processes.[2] Not only were prints crucial for circulating discoveries, but they also underpinned the drawing practices of Fellows as they would have learned to draw by studying printed images. The eye of the would-be draftsman had to be trained first to recognize a good image before embarking on making new images.

The development of judgment then was necessary for the communication of knowledge. By focusing on the move from an apprenticeship model to one based on self-education through print culture, this book builds on Pamela Smith's work which stressed the importance of physical experience in learning about nature.[3] It also builds on the work done by Susan Dackerman and her emphasis on prints as "tools or guides" as opposed to sources that were used "exclusively for predetermined facts."[4] Studying the images included in Robert Hooke's *Micrographia* (1665), Francis Willughby's *Ornithology* (1676), and the articles published in the *Philosophical Transactions* from 1665 to 1678 allows me to trace the myriad ways in which print intersected with the research agenda set by the Royal Society. Although two of the three texts chosen have single authors associated with them, the publication of both *Micrographia* and *Ornithology* were entwined with the activities of the Society as a whole.[5]

1 This phrase is taken from the title of Marie Boas Hall's book; she in turn takes it from the statutes of the Royal Society. Hall, *Promoting Experimental Learning*.

2 See for example: Stinjman, *Engraving and Etching* and Griffiths, *Print Before Photography*.

3 Smith, *Body of the Artisan*.

4 Dackerman, *Prints and the Pursuit of Knowledge*, 32.

5 As Sachiko Kusukawa has shown the only truly Society sponsored book from this time period was Francis Willughby's *Historia Piscium*. Kusukawa, "*Historia Piscium*," 179–197.

From its founding, the Royal Society worked to improve the state of knowledge about the natural world through experiment and observation. The Fellows of the Royal Society broke with past reliance on ancient opinions regarding the workings of nature and instead relied on their own experiences.[6] As Thomas Sprat wrote in his *History of the Royal Society* (1667):

> I shall lay it down, as their *Fundamental Law*, that whenever they could possibly get to *handle* the subject, the *Experiment* was still perform'd by some of the *Members* themselves. The want of this *exactness*, has very much diminish'd the credit of former *Naturalists*.[7]

Profession of this "Fundamental Law" set the Royal Society apart from scholars entrenched in the university system at the time and from many earlier naturalists, who as Sprat noted lacked the same commitment to "exactness."[8] Even if Sprat's *History* is more polemical than factual, this passage lays out the importance of ocular verification for the early Fellows.

One approach Fellows took to improve their own visual education was the active pursuit of a history of trades, which was based on Francis Bacon's call for improved methods of history in his *Advancement of Learning*.[9] In 1668 John Evelyn (1620–1706), a founding Fellow, published a translation of Roland Fréart's *Idée de la Perfection de la Peinture*, as part of the Society's larger project to create a history of trades.[10] The review of Evelyn's translation of Fréart's text in the *Philosophical Transactions*, in addition to giving an account of the book, encouraged readers to learn to draw, paint, and engrave for themselves stating that the book "will doubtless animate many among

6 Purver, *Royal Society*, esp. chap. 3, "A New System of Natural Philosophy II, The Royal Society." The literature on the place of experiment in the workings of the Royal Society is extensive. A few major titles include: Hall, *Promoting Experimental Learning*; Shapin and Schaffer, *Leviathan and the Air-Pump*; and Hunter, *Establishing the New Science*.

7 Sprat, *History of the Royal Society*, 83.

8 Purver notes that it should come as no surprise that the men who would eventually found the Royal Society began meeting at the University of Oxford and worked to enact reform from within the university system as this was the seat of the most concerted resistance to change. Purver, *Royal Society*, 63.

9 For an early account of the history of trades program see: Houghton, "The History of Trades," 33–60. For a more updated account see: Ochs, "The Royal Society," 129–158.

10 For Evelyn's participation in the history of trades programs, see: Hanson, *English Virtuoso*, 75–80. Evelyn, *Idea of the Perfection of Painting*. Evelyn made other contributions to the history of trades as well, for example: Evelyn, "A Letter...Concerning the Spanish Sembrador," 1055–1065; Evelyn, *Sculptura*; Evelyn, *Parallel of the Antient Architecture*. Sachiko Kusukawa has also discussed Evelyn's work on the Society's history of trades program: Kusukawa, "Early Royal Society," 360.

us to acquire a perfection in Pictures, Draughts and Chalcography, equal to our growth in all sorts of Optical Aydes, and to the fullness of our modern Discoveries."[11] The reviewer saw the study of artistic practices as a necessary adjunct to new discoveries. The study of art was put on a par with the use of telescopes, microscopes, and other optical instruments used in the study of nature. The importance of image making for the pursuits of Fellows of the Royal Society was stressed as the reviewer wondered: "what Art can be more helpful or more pleasing to a Philosophical Traveller, an Architect, and every ingenious Mechanician? All which must be lame without it."[12] The myriad lines of inquiry pursued by the Fellows were enhanced by a simultaneous attention to art. Within its role at the center of print production and the underlying emergence of accuracy that I trace in these pages, I argue that the engraver's burin, or graver, should be considered among the pantheon of instruments that fundamentally changed how nature was studied and subsequently understood in the seventeenth century.

Painting and sculpture were positioned as "the politest and noblest of Antient Arts, true, ingenuous, and claiming the Resemblance of Life, the Emulation of all Beauties, the fairest Records of all Appearances, Divine or Humane."[13] Yet, "claiming the Resemblance of Life" was not a simple task, nor is it clear from this passage what "resemblance" precisely meant. True and ingenuous hardly seem like synonyms. Are we to understand the truth-value of images to be based on the notion of their innocence and frankness? Or on their nobility and ties to the liberal arts?[14] This lack of guile and plenitude of honesty should immediately be called into question. In this period, ingenuous could also mean witty or cunning and "associations with dissimulation were always lurking beneath the surface."[15] Even if we accept the possibility that an image could be an unbiased "record of all Appearances," resemblance and emulation imply a gap between reality and representation. It is this gap in the mimetic functions of these images and others like them that this book explores. This gap is a productive space for

11 "Chalcography" was used in the period to refer specifically to engraving on copper plates, as in the title of Evelyn's history of engraving. *Oxford English Dictionary*, s.v. "Chalcography." "Account of Some Books," 785.

12 "Account of Some Books," 785.

13 "Account of Some Books," 785.

14 Although the *Oxford English Dictionary* states at the end of the entry for ingenuous that in the seventeenth century it was often misused in place of ingenious, the sources given in the *Oxford English Dictionary* for definition 2b of "ingenuous" ("Of animals or things: Of high or excellent quality or character; 'noble'.") are taken from the writing of John Evelyn.

15 Marr, et al., *Logodaedalus*, 213.

investigating what it meant to produce the effect of accuracy, and, further, how accuracy was read and used in the context of the early Royal Society of London.

"The Best in the World in this kind"

As has been shown by Pamela Smith and the authors involved in Lissa Robert's *The Mindful Hand* essay collection, to name but a few, scholars and artisans worked together to produce knowledge about the natural world.[16] *Engraving Accuracy* examines how these connections in Restoration London established protocols by which scholars and artisans could agree on the accuracy of an image. To explore the conditions in which producing images led to contrasting versions of accuracy in this period, I begin with two engravings by William Faithorne, a sought-after London engraver. The first is a portrait of Barbara Villiers, Lady Castemaine, and the second is the first plate in Samuel Collins' *A Systeme of Anatomy, Treating of the Body of Man, Beasts, Birds, Fish, Insects and Plants.*[17] (Fig. I.1 and I.2) Examining them together shows how the viewer's perception of technical skill and resemblance to life converged to create a concept of accuracy in scientific images for the Fellows of the Royal Society during its first thirty years, 1660–1690. The production of accuracy depended on the work of the hands and eyes of natural philosophers and engravers. This pair of images exemplify the rich visual culture of Restoration London and highlight the deep web of connections between artisans and natural philosophers that this book examines.

William Faithorne's portrait of Barbara Villiers, Lady Castlemaine, published in 1666, exemplifies the part of his oeuvre that is best known and written about most frequently.[18] Villiers was King Charles II's mistress beginning in 1660 soon after his return to London with his Restoration to the throne and ending roughly in 1672 with her displacement in favor of Louise de Kéroualle, Duchess of Portsmouth.[19] This engraving appeared

16 Smith, *Body of the Artisan* and Roberts, *The Mindful Hand.*

17 The spelling and punctuation of the primary sources have been maintained throughout. Unless otherwise noted, the italics are in the original.

18 For biographical information on Faithorne see: Griffiths, "Faithorne, William,"; Walpole, *Catalogue of Engravers*, 106–128; Godfrey, "William Faithorne," 208–213; Alexander, "Faithorne, Loggan Vandrebanc and White," 297–316. For a discussion of Faithorne's drawings see: Sloan, "Sir Hans Sloane's Pictures," 381–415.

19 MacLeod and Alexander, *Painted Ladies*, 116.

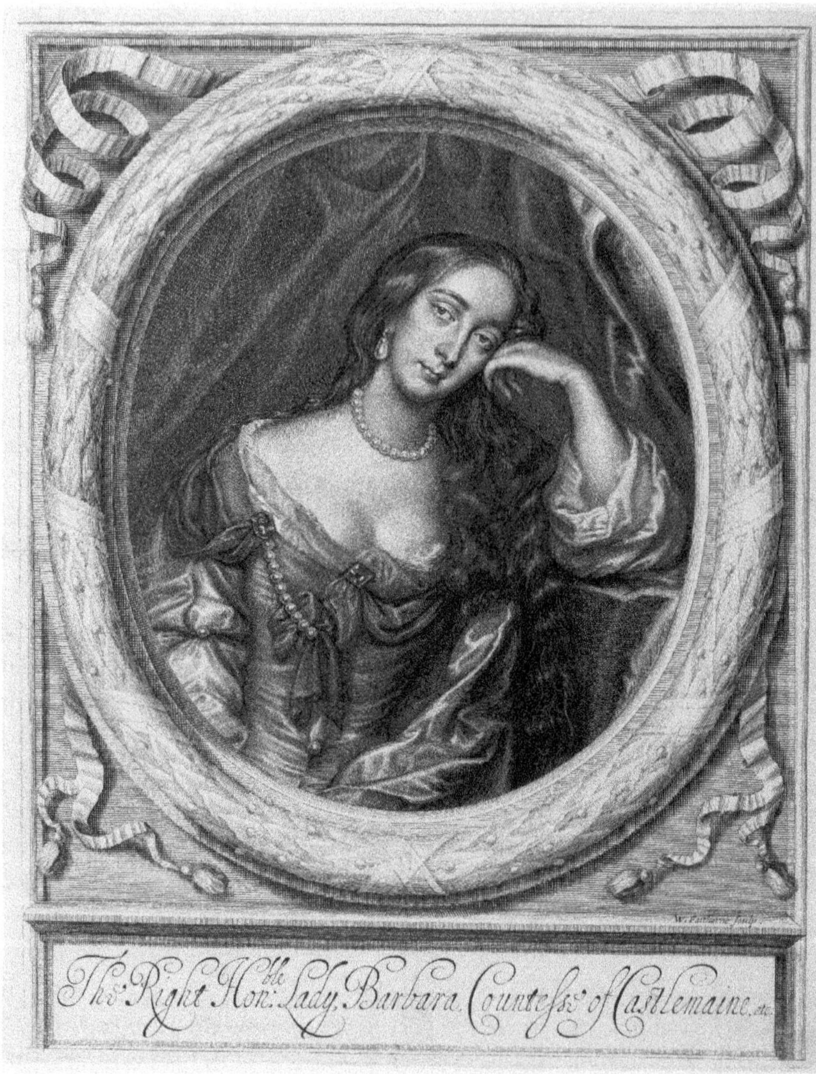

Figure I.1: William Faithorne, *Portrait of Barbara Villiers*, Engraving, 1666. Chazen Museum of Art, University of Wisconsin-Madison, Gift of Herbert Sewell, 1983.47.

on the London market at the height of Villiers's favor with the King and of the public's interest in her. Faithorne modeled his portrait on a painting by Sir Peter Lely of her as the penitent Magdalen.[20] This attribution was

20 Multiple copies of this painting exist, but Julia Marciari Alexander in her catalogue entry argues for the version now in the collection of the National Trust at Knole as the original and dates it to ca. 1662 (inv. no. 129855). MacLeod and Alexander, *Painted Ladies*, cat. no. 33, 118.

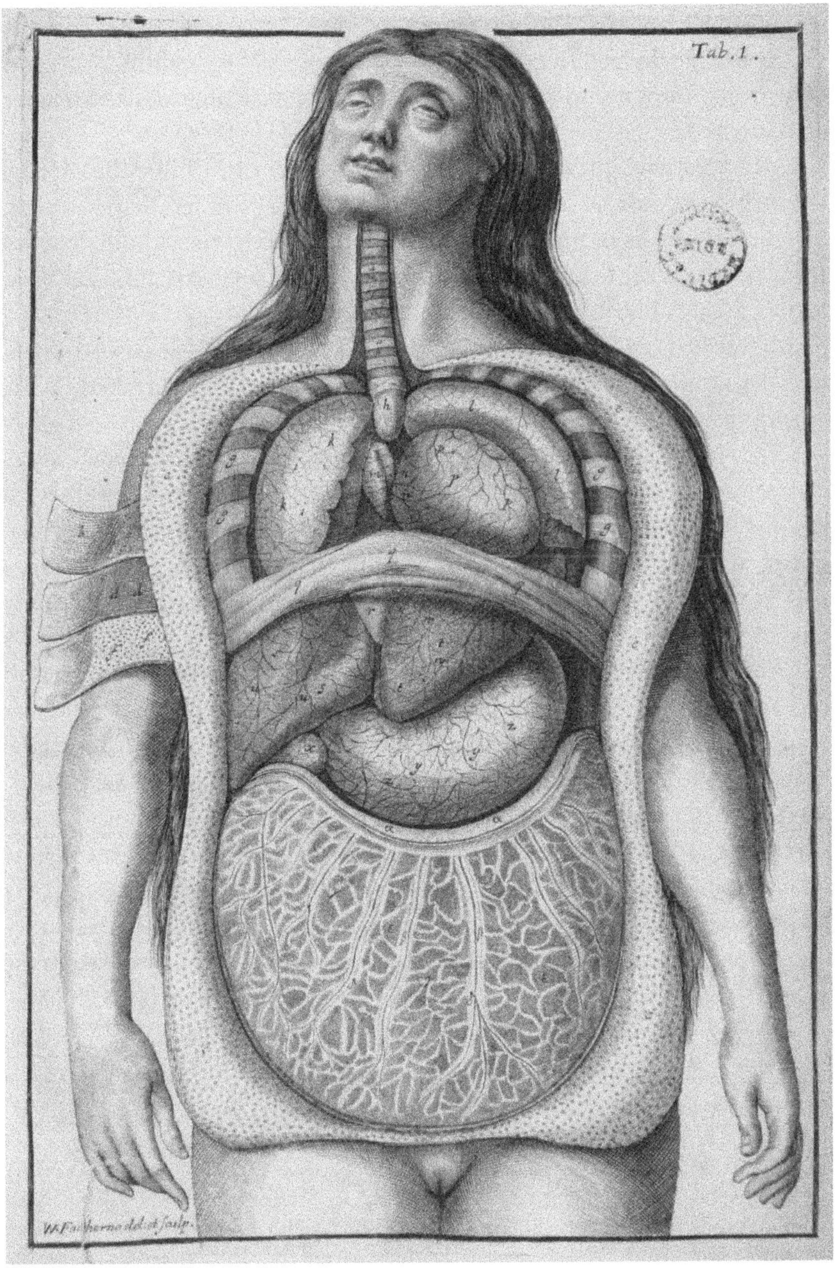

Figure I.2: William Faithorne, *A Human Body Opened*, Tab. 1. 2nd volume, Engraving, from Samuel Collins, *A Systeme of Anatomy, Treating of the Body of Man, Beasts, Birds, Fish, Insects and Plants*, 2 volumes (London: Printed by Thomas Newcomb, 1685). Courtesy of Ebling Library's Rare Books & Special Collections, University of Wisconsin-Madison, Historical Vault Oversize WZ 250 C713S 1685.

not, however, one that Faithorne made publicly. Unlike portraits where he clearly acknowledged his sources or others where he noted the piece was done from his own drawing of the sitter, in this one he maintained silence on its source.[21]

Faithorne was quite capable of capturing her luminous beauty by rendering the softness of her skin and the delicacy of her features with his burin. Lady Castlemaine gazes idly out at the viewer, acknowledging the viewer's presence without staring intently. She delicately rests her head on her graceful hand allowing the light to fall gently on her cheek and highlight the luster of her large pearl earring. Faithorne closely attended to every detail of her head: her eyelashes, the curve of her eyebrows, the plumpness of her lips, the ripples of her hair as it cascades down her shoulder and the overall smoothness of her skin. Her elegant costume is also rendered with great skill as the gemstones and pearls compete with the lustrous fabric for the viewer's attention. These minute details add up to a recognizable portrait of a renowned beauty even though Faithorne was working from another's painting, not his own observations of Lady Castlemaine.

Faithorne brought this same attention to the portraits he produced based on his own drawings of the sitter as, for instance, in his portrait of Robert Boyle.[22] (Fig. I.3) Boyle's hair gently flows over his shoulder and the lustrous drapery behind him glistens as the light hits it. Setting aside the generic difference between a portrait of a court beauty and a learned author, both portraits capture the details of the sitters' bodies and the material objects which signify their status. Boyle was not alone among the Fellows of the Royal Society to have his portrait drawn and engraved by Faithorne.[23] Many of the portraits he drew of Fellows became the frontispieces to their

21 For example, his portrait of John Ogilvy bears the text "P. Lilly Pinxit." An image of the Ogilvy portrait is available on the website of the National Gallery of Art, Washington, D.C., Ailsa Mellon Bruce Fund, 1974.116.16.

22 Bell and Poole list twenty-seven portraits that Faithorne signed "*ad vivum*." This list consists of "the signed and dated or named drawings and the engravings, where these are described as '*ad vivum*' or after originals by" Faithorne. Bell and Poole, "English Seventeenth-Century Portrait Drawings," 52–54. Faithorne's drawing of Boyle is in the collection of the Ashmolean Museum.

23 Faithorne is known to have drawn portraits from life of nine Fellows of the Royal Society. The Fellows who sat for Faithorne are: John Aubrey, Robert Boyle, Edmund Castell, Francis Glisson, Edmond King, Charles Leigh, Henry More, John Ray, and John Wallis. This list is based on Bell and Poole's list of portraits that were done from life. Bell and Poole, "English Seventeenth-Century Portrait Drawings," 53–54. Their list has been cross-referenced with the Royal Society's online database of Fellows.

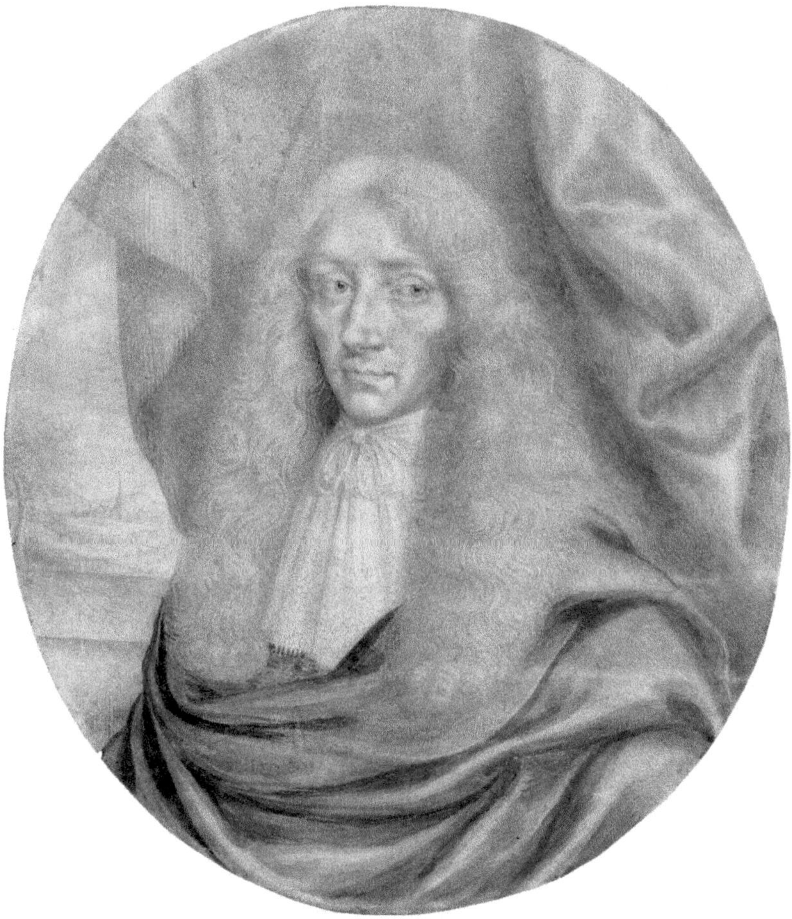

Figure I.3: William Faithorne, *Portrait of Robert Boyle*, Graphite on vellum, probably 1659–1665. ©Ashmolean Museum, WA.Suth.B.2.676.31.

books attesting to the value they placed on having a likeness of themselves prefaced to their published thoughts.[24]

Faithorne's ability to create a likeness of Barbara Villiers was noted in the period. Samuel Pepys, a Fellow of the Royal Society, wrote in his diary about seeing Faithorne's version of Lely's portrait and being compelled

24 See for instance Faithorne's portrait of John Ray, which was engraved by William Elder and used as the frontispiece for his *Synopsis Methodica* (1693), *Stirpium Europearum* [. . .] *Sylloge* (1694), *Wisdom of God* (1701), *and Three Discourses* (1732). The drawing for it is in the collection of the British Museum (Registration ID 1994, U.5). Croft-Murray and Paul Hulton, *Catalogue of British Drawings*, vol. 1 (London: The British Museum, 1960), 315–316.

to buy multiple copies. On December 1, 1666, Pepys was at Faithorne's "buying three of my Lady Castlemaynes heads, printed this day; which endeed is, as to the head, I think a very fine picture, and like her."[25] Pepys judged the value of the print by the similarity between the head as depicted in the engraving and as he knew the painting and the woman in life.[26] An avid collector of prints, Pepys amassed a huge collection during his lifetime, which included nearly 2,000 portrait prints.[27] This passage shows that he appreciated the image as both the work of a renowned artist and as a record of a lady's beauty.

Pepys was so taken by Faithorne's version of Lely's portrait that he attempted to buy the preparatory drawing a month before buying three copies of the print.[28] He recorded this visit to Faithorne's shop in his *Diary* and noted his preference for the "red chalke and other colours" that Faithorne used in creating his drawing after Lely's painting. However, Faithorne refused to sell the drawing as he needed to "keep it awhile to correct his Copper plate by," highlighting how the usefulness of the drawing to the process of creating a likeness outweighed the immediate rewards of selling his drawing.[29] Making the drawing served not simply as an intermediate step in the process, but also as a record of the painting against which Faithorne could "correct" his engraving. This note in Pepys's *Diary* shows that Faithorne took a great deal of care in creating an engraving and that a likeness could only be produced through a careful process of correction. Accuracy in this instance was constructed through skillful and careful actions.

As well as creating an image that resembled Lady Castlemaine, Faithorne was also capable of creating images that were useful to the communication of natural knowledge in the period. A physician looking to record a different type of likeness valued Faithorne's attention to careful working practices. In 1685 Samuel Collins, a noted anatomist and physician, published *A Systeme of Anatomy, Treating of the Body of*

25 *Diary of Samuel Pepys*, vol. 7, 393 (1 December 1666).

26 Pepys had seen Lely's original painting and vowed "I must have a copy." *Diary of Samuel Pepys*, vol. 3, 230 (20 October 1662). Pepys also saw Lady Castlemaine in the flesh in London in July 1660. MacLeod and Alexander go so far as to say that Pepys's "personal obsession with her bordered on the irrational." MacLeod and Alexander, *Painted Ladies*, 116. Among many other instances, Pepys records sitting near her at the theater. In this passage, he refers to her by her married name, Palmer: "...we seated ourselfs close by the King and Duke of Yorke and Madam Palmer (which was great content; and endeed, I can never enough admire her beauty)." *Diary of Samuel Pepys*, vol. 1, 174 (7 September 1661).

27 Latham, "Forward by the General Editor," ix.

28 *Diary of Samuel Pepys*, vol. 7, 359 (7 November 1666).

29 *Diary of Samuel Pepys*, vol. 7, 359 (7 November 1666).

Man, Beasts, Birds, Fish, Insects and Plants. Collins was physician to King Charles II and his *Systeme of Anatomy* was dedicated to Charles' successor, James II.[30] *A Systeme of Anatomy* stretches to 1,263 pages of text and was the product of many years of dissections of the full range of bodies listed in the title.

Beginning on the title page the importance of images to Collins' project is clear, the presence of the plates is noted immediately following the title of the work as the sub-title stated the book was "Illustrated with many Schemes."[31] The word "schemes," which in the period was used as a synonym for figure or plate, is printed in red with all capital letters.[32] These seventy-four plates are further described as "Consisting of Variety of Elegant Figures, drawn from the Life," and are collected at the end of the second volume of the book with their own dedication and preface. In the "Preface to the Tables," Collins stressed the "Care and Fidelity" he had taken in performing the dissections shown in the plates and established his own credentials as a capable anatomist.[33] After narrating his decision to make his private dissections public, Collins was not shy about heralding the qualifications of his choice of engraver; Faithorne was described as "an Excellent Artificer (if not the Best in the World in this kind)."[34] These qualifiers served both to honor Faithorne's work and to highlight Collins' good judgment. Many of the drawings made by Faithorne and used by him to produce the plates show clear signs of having been used for the engraving process as the outlines have been incised.[35] The seventy-four plates that Faithorne engraved for Collins were not based on someone else's design, like his portrait of Lady Castlemaine; rather they were based on Faithorne's own witnessing of Collins' dissections and were "Designed from the Life."[36]

Collins' "Preface to the Tables" set up his credentials and those of his artist and stressed that Faithorne was present at the dissections that he represented in print. As Steven Shapin and Simon Schaffer have shown in their work on the controversy between Robert Boyle and Thomas Hobbes, it

30 Cooper, "Collins, Samuel."

31 Collins, *Systeme of Anatomy*, title page.

32 Collins, *Systeme of Anatomy*, title page.

33 Collins, *Systeme of Anatomy*, "The Preface to the Tables," Sig ¶[1]1.

34 Collins, *Systeme of Anatomy*, "The Preface to the Tables," Sig ¶[1]1.

35 These drawings are preserved in the British Library as part of Add. MS 5260. The drawings are all laid down in an album so their versos cannot be examined to determine if chalk was applied to aid in the transfer process. However, the lines have been incised in the following folios: 1–12, 15–16, 18, 29, 32, 40, 45, 48, 50, 52–53, 75.

36 Collins, *Systeme of Anatomy*, "The Preface to the Tables," Sig ¶[1]1.

was not necessary to have many witnesses for the results of an experiment to be accepted; there just had to be a few credible witnesses.[37] Credibility in this instance is not only something gentlemen possess, but also artisans. Collins' reporting of Faithorne's presence at the dissections worked to add authority to Faithorne's engravings and to close the distance between line and life. This closeness, as was the case in the portrait of Lady Castlemaine, derived from Faithorne's working practices.

Looking at the first plate in *A Systeme of Anatomy* reveals a dramatic contrast with the portrait Lady Castlemaine. (Figure I.2) This plate depicts a woman with her torso opened to reveal her internal organs. In the lower left corner, Faithorne has signed the plate "W. Faithorne del: et sculp." indicating that he both drew the design for the plate and engraved it. Although this plate was not signed *ad vivum*, we are led to believe that Faithorne was working from his own observations of Collins' dissections based on Collins' assertion in the "Preface to the Tables" that Faithorne worked "from the Life." This was not the only instance of Faithorne being present at a dissection. Faithorne was noted as attending dissections of a rattlesnake by Edward Tyson at the Royal Society around this same time.[38] Edward Tyson was very involved in Collins' project as well as the dissections at the Royal Society.[39] Furthermore, an article in the December 1685 issue of the *Philosophical Transactions* by John Brown begins with a description of the image of a dissected liver "as it was accurately taken by Mr Faithorn [*sic*]."[40] The accuracy of Faithorne's drawing was based on his attendance at the dissection and the approbation of the other witnesses mentioned in the article, William Dawkins, William Briggs, and Edward Tyson, all of whom were Fellows of the College of Physicians.[41] The viewer of Collins's book is then led to assume that Faithorne brought the same careful hand and eye he used to create his celebrated portraits to his anatomical drawings and engravings.

The marked visual differences in these two images of women are not due to some lack of ability on Faithorne's part to render the human body but, rather, result precisely from his careful working habits and close observations of his objects of study. That is, the grey pallor of the woman and her oddly

37 Shapin and Schaffer, *Leviathan and the Air-Pump*, 55–60.

38 The account of Tyson's dissection was published in the *Philosophical Transactions*. Tyson, "Vipera Caudi-Sona Americana," 25–58.

39 Birch, *History of the Royal Society*, vol. 4, 176. Kim Sloan discusses these drawings in: Sloan, "Sir Hans Sloane's Pictures," 412.

40 Brown, "Remarkable Account," 1266.

41 Brown, "Remarkable Account," 1267.

elongated neck are not distorted renderings of a court beauty; they instead represent the face of a dead woman who was likely hanged before being dissected.[42] In contrast to Lady Castlemaine's lustrous skin and bright eyes, the tone of this anonymous woman's skin is built up with short, overlapping stokes that give her a greyish pallor. Her eyes roll back slightly in their sockets and look upward in an unfocused manner. The deep bags under her eyes are highlighted by the angle at which her head sits. Faithorne was not using the same conventions to depict these two female bodies.

These conventions were both situational and formal: situational in that wealthy women were often accompanied by visual indicators of their status, such as the lustrous fabrics and jewels that adorn Lady Castlemaine. The formal differences between these two images of women extend beyond their very different compositions. The lines Faithorne used to indicate the texture of Lady Castlemaine's skin are long and flowing, whereas those delineating the tone of the cadaver's skin are coarser and her face is more heavily shaded. Faithorne's careful working habits yielded very different images. Both were understood by their audiences to achieve the effect of accuracy. Scholars and artisans worked to meet the expectations of their readers/viewers by creating images that acknowledged the audience's shared experiences of the three-dimensional world being portrayed with the engraved line.

Faithorne's cadaver looks neither like his portrait of Lady Castelmaine, nor like previously published anatomical images. What sets it apart is an attention to conforming to the shared experiences of the artist and the reader. Andreas Vesalius's images of dissected women in his *De Humani Corporis Fabrica* did not resemble recently hanged humans. (Fig. I.4) Instead the dissected anatomy of a woman was placed into the torso of a classical statue. Although Vesalius's anatomical investigations greatly improved scholars' understandings of the inner workings of the human body, the images of female bodies were not showing anatomists what they might expect to see while dissecting.[43] What Vesalius' plates provided were fantasies: headless, armless, legless torsos were containers for representations of

42 By the end of the seventeenth century, there was a long-standing precedent for using the bodies of condemned criminals for anatomical dissections, as this was common practice as early as 1300. Andreas Vesalius tells his readers that the dissected woman described in his book, *De Humani Corposis Fabrica*, was a criminal and that she had been hanged. For an in-depth discussion of Vesalius's dissection see, Park, *Secrets of Women*, 207–215. In the case of England, Ruth Richardson writes, "Since Henry VIII's time, the sole legal source for corpses for dissections had been the gallows." Richardson, *Death, Dissection, and the Destitute*, xv.

43 For an analysis of Vesalius's use of images see: Kusukawa, *Picturing the Book of Nature*.

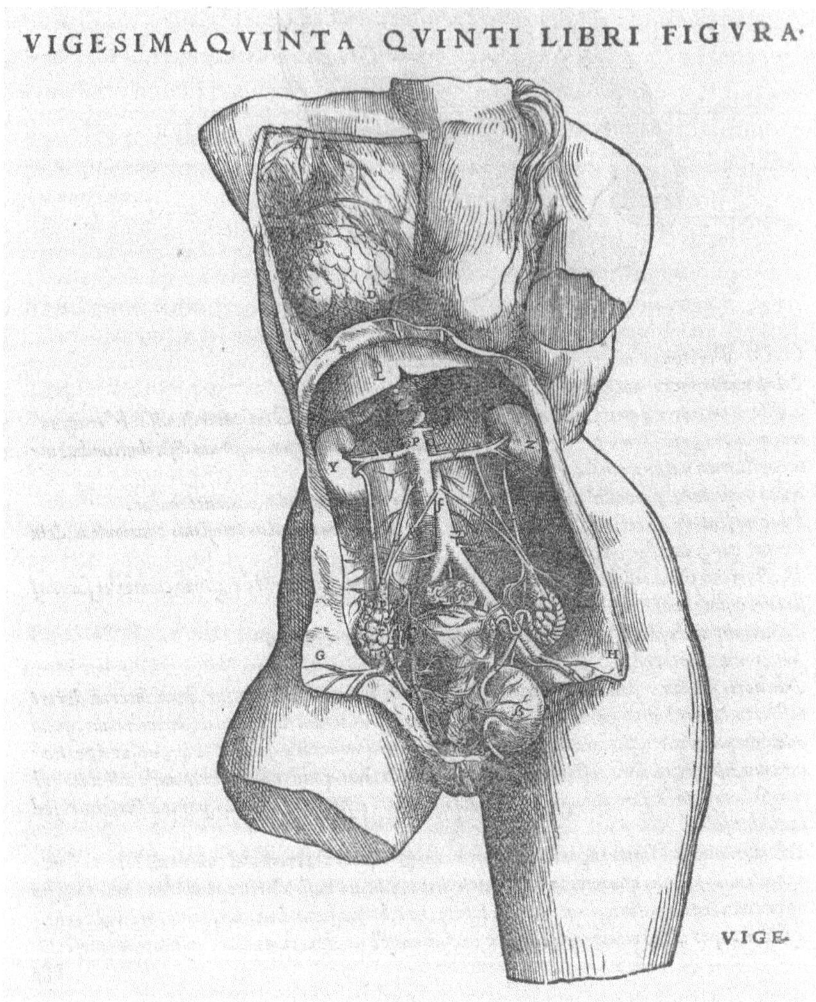

Figure I.4: Jan Stephan van Calcar, *Anatomy of Female Torso*, page 478, Woodcut, from Andreas Vesalius, *De Humani Corporis Fabrica* (Basel: Oporinus, 1543). Basel University Library via e-rara.ch.

internal organs, not dissected bodies. While Faithorne's image of a female cadaver was clearly edited to clarify the relationship of the internal organs to one another, the body that contains them more closely resembles what an anatomist might expect to find lying on the table before him.

At the beginning of the seventeenth century, Guilio Casseri (1552–1616) published his comparative study of the sensory organs and the parts of the body being dissected now appear connected to actual heads.[44] (Fig. I.5)

44 Casseri, *Pentaestheseion*.

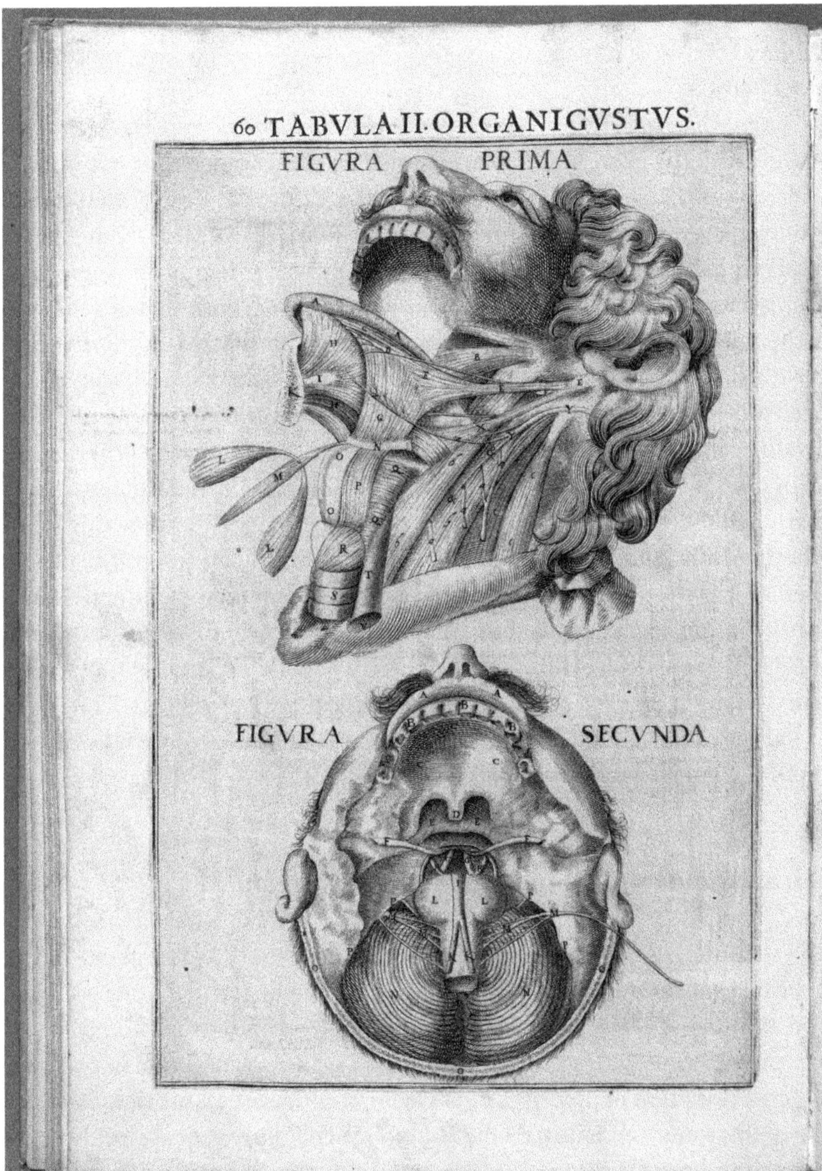

Figure I.5: Anonymous, *Oragani Gustus (Taste)*, plate two, page 60, Engraving, from Gulio Casseri, *Pentaestheseion, hoc est De qvinqve sensibvs liber* (Venice: Apud Nicolaum Misserinum, 1609). The Huntington Library, San Marino, California, RB 631661.

However, the heads appear to be perfect. Each hair is perfectly placed. The ends of the moustaches curl up neatly. These images present highly detailed renderings of the anatomy of the sensory organs attached to the heads of perfectly coiffed Italian courtiers. Instead of resembling a court beauty, a classical statue or a handsome young man, Faithorne's image looks like a dead body.

Rather than referring to the life of the object of study, "from the life" refers to the life of the observer. Where Samuel Pepys was comparing Faithorne's portrait with his own experience of seeing Villiers at Court, anatomists looking at Collins's book would have compared Faithorne's plates with dissections they had witnessed. That is, they were judging the image not against their knowledge of a court beauty, but rather against that of a cadaver. Although these images *look* quite different, they both derive authority from the illusion that they are close to their referents, which in both cases have been altered to conform to standards of representation, the conventions of beauty in one case and those of anatomical dissection on the other.

In conforming to these two distinct sets of standards, Faithorne was further showing viewers that he was "the Best in the World in this kind" whether "this kind" was a portrait or an anatomical illustration. The viewer then needed to take the context in which the image was encountered into consideration when examining it. Accuracy in the seventeenth century then was not transhistorical or transdisciplinary but an illusion that functioned in a particular context and required the active work of viewers. Their own visual education and understandings of the image making process permitted them to analyze and accept this illusion.

"An accurate impression is in far higher esteem"

The acceptance of this illusion was predicated on the emergence of the habitus that developed in seventeenth-century England among scholars and artisans. As Pierre Bourdieu defined it, "The habitus is…laid down in each agent by his earliest upbringing, which is the precondition not only for the co-ordination of practices but also for practices of co-ordination, since the corrections and adjustments the agents themselves consciously carry out presuppose their mastery of a common code."[45] Sachiko Kusukawa's extensive research project on the role of images in the workings of the early Royal Society has uncovered myriad connections between Fellows of the Royal Society and artisans.[46] Their shared emphasis on visual practices

45 Bourdieu, *Outline of a Theory*, 81.

46 Kusukawa, "Early Royal Society," 350–394; Henderson, "Robert Hooke," 395–434; Reinhart, "Richard Waller," 435–484.

extended to the Continental correspondents of the Royal Society, such as Antoni van Leeuwenhoek as Sietske Fransen has shown.[47]

The common code shared by artisans and scholars was built upon the technologies of witnessing described by Steven Shapin and Simon Schaffer as well as the "tyranny of the rule," as William Ivins called the style of engraving that developed for reproductive prints in the sixteenth century.[48] For Ivins the "exactly repeatable pictorial statement" of the engravers whose primary output was reproductions of paintings marked the nadir of the printmaking and produced:

> what geometers call the 'net of rationality', a geometrical construction that catches all the so-called rational points and lines in space but completely misses the infinitely more numerous and interesting irrational points and lines in space. The effect of these rationalized webs on both vision and visual statement was a tyranny, that, before it was broken up, had subjected large parts of the world to the rule of a blinding and methodically blighting visual common sense.[49]

It is precisely this "blighting visual common sense" that allowed natural philosophers to accept the accuracy of engravings.[50] This tyrannical rule was how scholars learned to see and depict the three-dimensional world around them.

The images considered in what follows are representations of objects. They attempt to depict what individual scholars saw when they looked at the world around them. Whether they were looking through a microscope, a telescope, or with the naked eye, they were trying to understand nature through observation. How Fellows of the Royal Society made sense of experiments was similarly dependent on visual evidence. By and large, Fellows were trying to communicate their understandings of objects, not spaces. These images created the illusion of three-dimensional objects and were not wedded to the rules of Albertian linear perspective.[51] In this, I am following Lorraine

47 Fransen, "Antoni van Leeuwenhoek," 485–544.

48 Shapin and Schaffer, *Leviathan and the Air-Pump*. Ivins, *Prints and Visual Communication*, chap. 4: "The Tyranny of the Rule: The Seventeenth and Eighteenth Centuries," 71–92.

49 Ivins, *Prints and Visual Communication*, 70.

50 Art historians have worked to break down Ivins' analysis of this development particularly in the Italian context. See for example: Bury, *The Print in Italy,* and Lincoln, *The Invention of the Italian Renaissance Printmaker.*

51 Pamela Smith nicely summarized the historiographical thread which focused on the links between "science" and linear perspective: Smith, "Art, Science, and Visual Culture," 88–89.

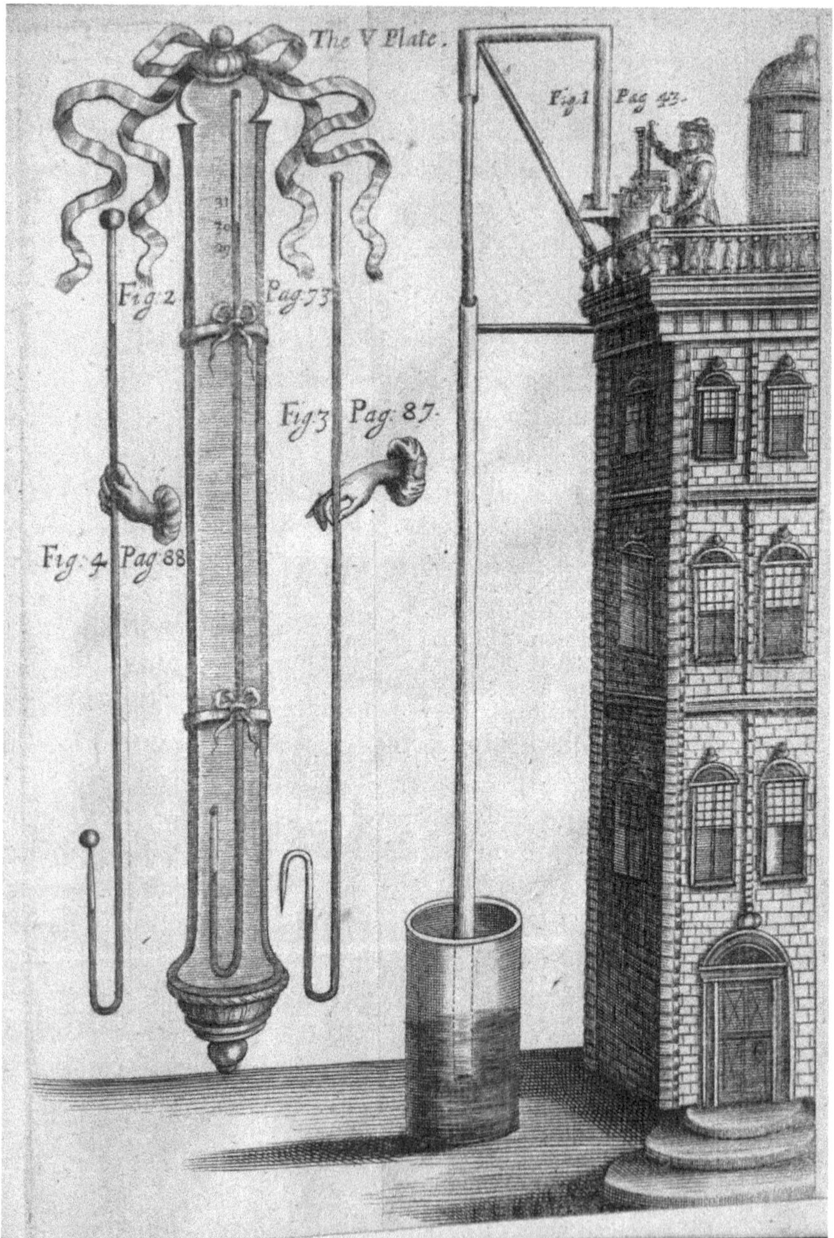

Figure I.6: Anonymous, *Plate five*, Engraving, from Robert Boyle, *A continuation of new experiments physico-mechanical* (Oxford: Printed by Henry Hall printer to the University, for Richard Davis, 1669). By courtesy of the Department of Special Collections, Memorial Library, University of Wisconsin-Madison, Duveen D263.

Daston's definition of an epistemic image as "one made with the intent not only of depicting the object of scientific inquiry but also of replacing it."[52] As such these engravings provided readers access to other scholar's objects of study.

Instead of presenting images of laboratories that readers could *inhabit*, as in Albertian conceptions of space, Fellows presented images of objects that readers could *use*. My emphasis on the functionality of these images is similar to Susan Dackerman's argument "that prints were also important tools in the processes of inquiry, fabricated and activated by their makers to serve as dynamic matrices for argumentation and persuasion."[53] In a similar vein, Anthony Griffiths has noted that the Fellows of the Royal Society needed a "medium that could be studied outside the laboratory."[54] For instance, when Robert Boyle provided further details about how to construct and use the air pump, his images did not convey a perfect sense of habitable buildings set in an English landscape.[55] (Fig. I.6) What is clear in figure one of the plate is the relationship of the parts of the instrument and the building. That is, Boyle and his artist emphasized how the instrument works and connects to the structure. Further, this is not a rendering of a building in a landscape because the rest of the space on the plate is given over to representations of other instruments. These, too, are concerned with how the parts relate to the whole and how to manipulate the objects to conduct experiments. The "where" of the laboratory gives way to the "what" of the instrument and its components.

Focusing on observational and experimental practices shifts the focus away from an emphasis on linear perspective as the primary means of creating a "true" representation of the three-dimensional world. Even those critics such as Lyle Massey who question the insistence on perspective as the dominant feature in images of the three-dimensional world, continue to reinforce the dominant position of perspective in the literature on art and science.[56] Fellows of the Royal Society learned to produce and consume images that projected three-dimensional objects onto paper using engraved lines. The accuracy of these images was dependent on makers' and users' shared participation in a visual culture informed by the engraved line.

With the end of the Civil War and the subsequent restoration of the monarchy, London was awash in printed images. Royalist artists who had

52 Daston, "Epistemic Images," 17. For an overview of the, at times divergent, definitions of "epistemic images," see: Marr, "Knowing Images," 1006–1008.

53 Dackerman, *Prints and the Pursuit of Knowledge*, 33.

54 Griffiths, *Print in Stuart Britain*, 267.

55 Boyle, *Continuation.*

56 Massey, *Picturing Space.*

fled to the Continent during the war returned to their trades invigorated by their exposure to the thriving print market in Paris and the Low Countries. At the same time there was an influx of Continental artists who emigrated to London. As a result, Antony Griffiths has identified this moment as "a watershed in British history" particularly with respect to printmaking.[57] These artists brought with them an awareness of and competency in engraving after others' designs. They were some of the "busy spiders of the exactly repeatable pictorial statement" to use William Ivins' phrase.[58]

The engraver's burin was a key instrument in the production of accuracy. Despite the additional cost and the need for a second press to include an engraving rather than a woodcut in a book, by the middle of the seventeenth century most printed images used in the pursuit of natural knowledge were engravings and etchings.[59] Intaglio prints were valued for their fine lines. In an undated essay on "chalcography," John Beale, a Fellow of the Royal Society, noted his preference for them stating that they "hath already gotten so much credit, that to all ingenious person an accurate impression is in far higher esteem, than the most gaudy paintings of vulgar note."[60] Precise lines from a burin were valued over the bright colors applied by the brush. Beale's comment followed the lines of the argument Giorgio Varsari made regarding the value of Florentine *disegno* in contrast to Venetian *colore*, where *disegno* was aligned with the more intellectual aspects of drawing an image with line and *colore* with the production of an image through the layering of paint.[61] Richard Haydocke's translation of Lomazzo's *Trattato dell'arte della pittura* brought the English-speaking world into direct contact with the Italian debate.[62] Through this allusion to the Italian debate, Beale imbued the burin with a sense of rationality as opposed to the more emotional work of the brush thereby making it easier to accept the images produced by it as serving the goals of natural philosophers.

By combining engraving and etching, printmakers rendered the minute detail in a fly's eye, the iridescence of a hummingbird's feathers, a comet's broad tail passing through the solar system, among myriad other objects of study for natural historians and natural philosophers. Although they are

57 Griffiths, *Print in Stuart Britain*, 23.

58 Ivins, *Prints and Visual Communication*, 70.

59 For a discussion of the relationships between letter press printers and copperplate printers, see: Gaskell, "Printing House," 213–251.

60 Beale, "Of Chalcography on humble motion," Archives of the Royal Society, CLP/2/22, quoted in Hunter, *Wicked Intelligence*, 128–129.

61 Pace, "Disegno e colore."

62 Pace, "Disegno e colore."

now often discussed and studied separately, in the early modern period etching and engraving were used simultaneously in many images and both were referred to as "graving."[63] Whether the burin was cutting the line directly into the copper or an acid bath was doing that work, lines cut into copper were essential to the circulation of knowledge. Similarly, Alina Payne has argued that perspective and painting worked as instruments alongside the telescope.[64] Much like how the telescope "made the act of seeing physically apprehensible," the burin made seeing apprehensible for others beside the viewer.[65] Producing accuracy with lines required careful actions and excellent tools.

In the seventeenth century, as Steven Shapin has argued, both standards and the truth were socially contingent concepts based on agreement between gentlemen, and it is at this very moment that the term accuracy comes into use.[66] In the context of histories of time, Stefan Hanß has pointed out the ways in which "historiography's fetish of accuracy" has imposed modern obsessions onto historical actors.[67] This book is not bringing a modern sense of accuracy to the seventeenth century, but rather is unpacking what the word meant for the early Fellows of the Royal Society.

The *Oxford English Dictionary* defines "accurate" as "conforming exactly with the truth or with a given standard."[68] By presenting standard and truth as if they were equivalent, the very definition of accurate elides the difference between a standard and the truth. There is, however, a significant and productive gap between the two. A standard, further, is "an authoritative or recognized exemplar of correctness, perfection, or some definite degree of any quality."[69] "Accurate" is here tied to authority and exemplarity implying that consensus has been achieved within a group. The truth, however, is defined as "conformity with fact; agreement with

63 My research indicates that engraving was used as a generic term in the period and engraving and etching were seen as complementary processes and were often used together in a single image. For example, William Faithorne places "graveing" before etching in the title of his engraving manual, *The Art of Graveing and Etching*, even though three quarters of the text is dedicated to the chemical intricacies that must be mastered with etching. In addition, the section on etching is titled "The Art of Graving with Aqua fortis." Faithorne, *Art of Graveing*, 30. Throughout this book, in keeping with period usage, I will use "engraving" as the generic term for the intaglio processes prevalent in the seventeenth century.

64 Payne, *Vision and its Instruments*, 2–3.

65 Payne, *Vision and its Instruments*, 2.

66 Shapin, *Social History of Truth*.

67 Hanß, "Fetish of Accuracy," 267.

68 *Oxford English Dictionary*, s.v. "accurate," definition 3.

69 *Oxford English Dictionary*, s.v. "standard," definition II, 10, a.

reality; accuracy or correctness in a of statement, thought, etc." and is often assumed to exist outside of the social.[70] While the modern definition discussed above focusses on conformity, the early modern definition, noted by the *OED* as being obsolete, was "executed with care, careful."[71] The *OED* also notes that this definition is "now merged with" the definition discussed above. This book traces how the careful production of images was a key step in this semantic shift.

It is telling that the word accurate and its derivatives came into the English language around the same time as the founding of the Royal Society and that Fellow's writings are used as evidence of the words' occurrences.[72] When the terms entered the English language, they were still close to their Latin root, "curare" (to care for), relating to being performed with care.[73] This points to an increase in the importance of careful actions for the production of experimentally based knowledge. The same emphasis on careful action also emerges in the writings of artisans of the time. This shared interest is key to the habitus explored in this book. In the context of the visual culture of the early Royal Society, images carved into copper by artisans were produced through a series of careful actions, on the part of the artisans and the Fellows, and it was these actions that allowed the images to conform to a standard and ultimately be considered accurate. Accuracy then should be understood as a cultural construct.

Accuracy provided a fundamental, central, and transferrable criterion that grounded the emerging disciplines of natural philosophy. In the context of the Royal Society of London, these developments grew out of their Baconian program, and consequently, I take accuracy to be one of the wellsprings of modern science. What we now recognize as science in many ways grew from the drive for accuracy.[74] At its crux, this book explores how printed images achieved authority in the context of the early Royal Society. The

70 *Oxford English Dictionary*, s.v. "truth," definition III, 10, a.

71 *Oxford English Dictionary*, s.v. "accurate," definition 1.

72 *Oxford English Dictionary*, s.v. "accurate," "accuracy," "accurately," and "accurateness."

73 *Oxford English Dictionary*, s.v. "accurate," etymology.

74 In this book, I do not want to enter into the overworked territory of whether or not there was a Scientific Revolution, nor do I intend to rehearse the historiography of the topic. Lorraine Daston and Katharine Park in their introduction to the Cambridge History of Science volume on *Early Modern Science* chose to forego using the phrase at all and instead use "early modern science" to demarcate the period under consideration in their volume. They go so far as to say that the "Scientific Revolution" is in fact a mythology. Daston and Park, eds., *Cambridge History of Science*, vol. 3, 15. Instead, I want to press on the assertion that collaborations between artisans and scholars affected the nature of research in the seventeenth century. Lissa Roberts and Simon Schaffer, furthermore, assert that the rhetoric of revolutions is based, in part, on the

careful processes of carving, which visually signify accuracy, were coupled with detailed discussions of method to secure printed images' immediate and urgent relationships to knowledge.

"Each Judgement of his Eye"

Fellows sought to minimize the distance between themselves and their objects of study. Abraham Cowley nicely captured this commitment to closeness with their object of study in his lengthy ode "To the Royal Society," which was included at the beginning of Sprat's *History of the Royal Society*.

> Who to the life an exact Piece would make,
> Must not from Others Work a Copy take;
>   No, not from *Rubens* or *Vandike*;
> Much less content himself to make it like
> Th'Ideas and the Images which ly
> In his own Fancy, or his Memory.
>   No, he before his sight must place
>   The Natural and Living Face;
>   The real Object must command
> Each Judgment of his Eye, and Motion of his Hand.[75]

This passage summarized the dominant strands and concerns of contemporary art education while subverting them to prioritize first-hand observation. The traditional studio practice of copying masters' works is rejected, as is the controversial practice of relying on the imagination.[76] Instead, the close examination of "the real Object" was paramount. Making "an exact Piece" was dependent on training both the hand and the eye.

Direct observation and the creation of a visual record of those observations were essential to Fellow's view of their collective project. The emphasis on individualized, first-hand observation set the Fellows apart from earlier

assumption that the mind and the hand, the scholar and the artisan, are distinct. Roberts and Schaffer, "Preface," in *The Mindful Hand*, xiii.

75 Cowley, "To the Royal Society," in Sprat, *History of the Royal Society*, B2r. I have tried to faithfully capture the original mise-en-page of the printing of Cowley's poem.

76 For a discussion of the religious concerns regarding a reliance on the imagination for inspiration see: Parshall, "Graphic Knowledge," 393–410.

Continental uses of the terms "*counterfacta*," "*naer het leven*," and "*ad vivum*." In writing about the use of the term "*counterfacta*" and its vernacular translations, Peter Parshall argues that the term did not necessarily imply that the artist was working from the life and uses the example of Dürer's drawing and later prints of the rhinoceros as evidence.[77] With regard to sixteenth- and early seventeenth-century botany, Claudia Swan has written that "naer het leven" served as an "internationally valid password" that signified general working practices and that the "documentary value of such images does not depend on individual authorship."[78] Sachiko Kusukawa surveyed the use of *ad vivum* in Latinate natural history from the mid-sixteenth century and found that "ad vivum signaled an image that a learned audience was expected to be able to recognize and understand" rather than signifying a singular working method that relied on direct observation.[79] In the context of the Royal Society, there are also cases where images were not "from the life" but the authors use this phrase to describe their process.[80] What is essential to note here is that even if it was not the case that Fellows were using images they had drawn, there was an emphasis put on the importance of being able to draw and on drawing from nature. The methods propounded by the Royal Society as a corporate body were taken up and practiced by the Fellows as the following chapters on Robert Hooke, John Ray and Francis Willughby, and Henry Oldenburg and his correspondents will show.

Members of the Royal Society included intaglio printed images in their books as stand-ins for the objects depicted despite the great distance between the object and a representation of it. These images need to be considered in a social context because like the "fact" they represented mediated knowledge. Barbara Shapiro writes that the "fact" did not begin as something that existed in nature, but instead it originated in the field of law.[81] In addition, Mary Poovey draws attention to dual meaning of the "modern fact" as something both outside of theory and imbedded in a social context and identifies the early Royal Society as a key site for understanding this duality.[82] In formal terms, an engraving translates an individual's experience of a full-color, three-dimensional object into a black and white re-presentation created with line. The observer's preconceived notions regarding the natural world

77 Parshall, "*Imago Contrafacta*," 554–579.

78 Swan, "*Ad vivum*," 364, 363.

79 Kusukawa, "*Ad vivum*," 89–121.

80 Kusukawa, "*Historia Piscium*," 186.

81 Shapiro, *Culture of Fact*, 105.

82 Poovey, *History of the Modern Fact*, chap. 3, "The Political Anatomy of the Economy: English Science and Irish Land," 92–143.

also mediate the knowledge presented in these images. The reality of the difference between a court beauty and a cadaver informed the creation of those images. By unpacking what accuracy meant in the period, this book contributes furthermore to previous scholarship both on "truth" and "fact" as socially constructed terms by showing that accuracy was also defined based on consensus.

This notion of accuracy is distinct from the work of Lorraine Daston and Peter Galison on the development of scientific objectivity and their concept of "truth-to-nature."[83] While they are concerned with the nineteenth-century epistemic shifts that resulted in objectivity assuming prominence in scientific image-making, I am interested in an earlier moment: I ask what the relationship between knowledge and the processes that caused it to be recognized were before natural philosophers and natural historians sought to distance themselves from the knowledge they produced. As Daston and Galison rightly note, objectivity was not a concern in the seventeenth century.[84] Furthermore, the images under consideration pre-date the set of concerns Daston and Galison identify as "truth-to-nature."[85] Instead, the images and practices examined in this book relate to what Daston and Galison describe as "collections of anomalies."[86] I argue that accuracy was the goal of those producing images for consumption by the Fellows of the Royal Society and their correspondents. Accuracy then is to truth-to-nature as truth-to-nature is to objectivity. These three terms define the goals of three centuries of scientific image-making and this book explores what accuracy meant for the Fellows of the Royal Society in the middle of the seventeenth century.

Throughout its four chapters, this book avoids the assumption that an image that was useful to the Royal Society simply looked accurate according to an implicit or explicit standard. Instead, this book examines the collective practices that stemmed from the habitus shared by artisans and scholars to show how they produced an effect of accuracy through social conditions and technical innovation. "Accuracy" is not a neutral or natural term, rather it results from a mediated set of effects that this book excavates. The collective practices discussed in these chapters include: the disciplining of artisans and observers/experimenters; the use of the same tools to achieve accuracy as to teach accurate actions; the textual presentation of an image;

83 Daston and Galison, *Objectivity*.

84 Daston and Galison, *Objectivity*, 27–35.

85 Daston and Galison, *Objectivity*, 55–113.

86 Daston and Galison, *Objectivity*, 67.

the inclusion of visual references to other previously printed images; and the role of consensus-building through circulation. This book argues that it was an adherence to these collective practices and not the giftedness of individual artisans or observers that created an effect of accuracy.

"Examining it according to my usual manner"

Asking what accuracy meant in terms of intaglio printed images produced for the Royal Society yields a deeper understanding of the imbricated roles played by the engraver and the observer or experimenter. A well-trained eye was essential to this execution of these actions. As Cowley put it nature "must command each Judgment of his [Philosophy's] Eye, and Motion of his Hand."[87] The eye and the hand then had to work together "for the Improving of Natural Knowledge."[88] Accuracy is here understood to be the combined effect of good judgment in selecting a referent and framing the image and careful actions in drawing and carving the image. This interleaved set of effects, both in production and consumption, contributed to the construction of signs that could be trusted.

Images produced by and for the Fellows of the Royal Society are distinct from those produced previously and in other European intellectual centers. Their differences are linked to the intellectual goals of the Royal Society and to the intimate ties Fellows had to the production of their images. Robert Hooke's name has become synonymous with the visual culture of the early Royal Society, and as will be discussed in detail in chapter two this was also true in the period. In Hooke's *Micrographia*, the illustration of the head of a fly fills the entire large plate and spills out of the confines of the book.[89] (Fig. I.7) The facets of the fly's compound eyes are clearly delineated as well as the fine details of the mouth. After describing how he prepared his specimen for viewing, Hooke narrated his procedures for looking as well as his results: "Then examining it according to my usual manner, by varying the degrees of light, and altering its position to each kinde of light, I drew that representation of it which is delineated in the 24.

87 Cowley, "To the Royal Society," in Sprat, *History of the Royal Society*, B2r.

88 This phrase is borrowed from the title of Thomas Sprat's *The History of the Royal Society of London for the Improving of Natural Knowledge*.

89 Hooke, *Micrographia*, Scheme 24. The sheet for the fly's head is 300 mm x 265 mm. These measurements are for the Burndy Library copy at the Huntington Library (752346). In this copy, the sheet has been trimmed within the plate mark at the top, and the right-hand plate mark is bound into the center of the volume.

Scheme, and found these things to be as plain and evident, as notable and pleasant."[90] Aside from the shadow cast by its seemingly enormous head, nothing else is included on the plate. By detailing his working methods and providing his readers with an image of what they might expect to see if they too proceeded in his "usual manner," Hooke produced the illusion that accuracy emerged from his working practices, rather than through an artifice in which his readers actively participated.

The circumscribed view of Hooke's fly, a view that in other plates corresponded to the scope of his view finder, provided readers with a marked contrast to the rich allegory that accompanied the first published images created with the help of a microscope.[91] In 1625 members of the Accademia dei Lincei published a broadsheet, known as the *Melissographia*, depicting three bees shown in microscopic detail along with close-ups of a number of dissected parts.[92] (Fig. I.8) This large-scale image was published as part of a year-long celebration of the election of Pope Urban VIII, a member of the Barberini family, whose family crest is three bees arranged in a trigon.[93] The trigon in the Lincei's broadsheet is formed of three views of a single bee: seen from above, below, and the side. Together these three views provided a highly magnified, complete view of the structure of the bee. The trigon is framed on the left and the right by bay branches, which house putti holding the symbols of the papacy; the putti on the left raises the papal tiara above his head and the one on the right is entwined in the papal keys. The bay branches are joined at the top of the sheet with a banner proclaiming the Lincei's devotion to Urban VIII, their patron.

While the upper two thirds of the sheet are given over to the whole bee and a display of devotion to the Barberini Pope, the lower third presents more detailed views of individual parts of the bee. A trompe l'oeil scroll stretches and curls across the bottom of the image and enlarged details of the body of bee unfold before our eyes. As the left-hand section of the scroll unfurls from around the ends of the bay branches the viewer is presented with two close-ups of the full head of the bee and a further detail of the parts of the mouth. The head is shown from the front and the side; the side view gives a magnified view of the structure of the bee's mouth as the details of this are lost in the bay branches as the bee seen from the side stretches its proboscis

90 Hooke, *Micrographia*, 175.

91 For detailed studies of the early history of the microscope see: Wilson, *Invisible World* and Fournier, *Fabric of Life*.

92 For a detailed study of the Lincean Academy see: Freedberg, *Eye of the Lynx*.

93 The sheet measures 41.6 by 30.7 cm. These are the dimensions listed in Freedberg, *Eye of the Lynx*, 160.

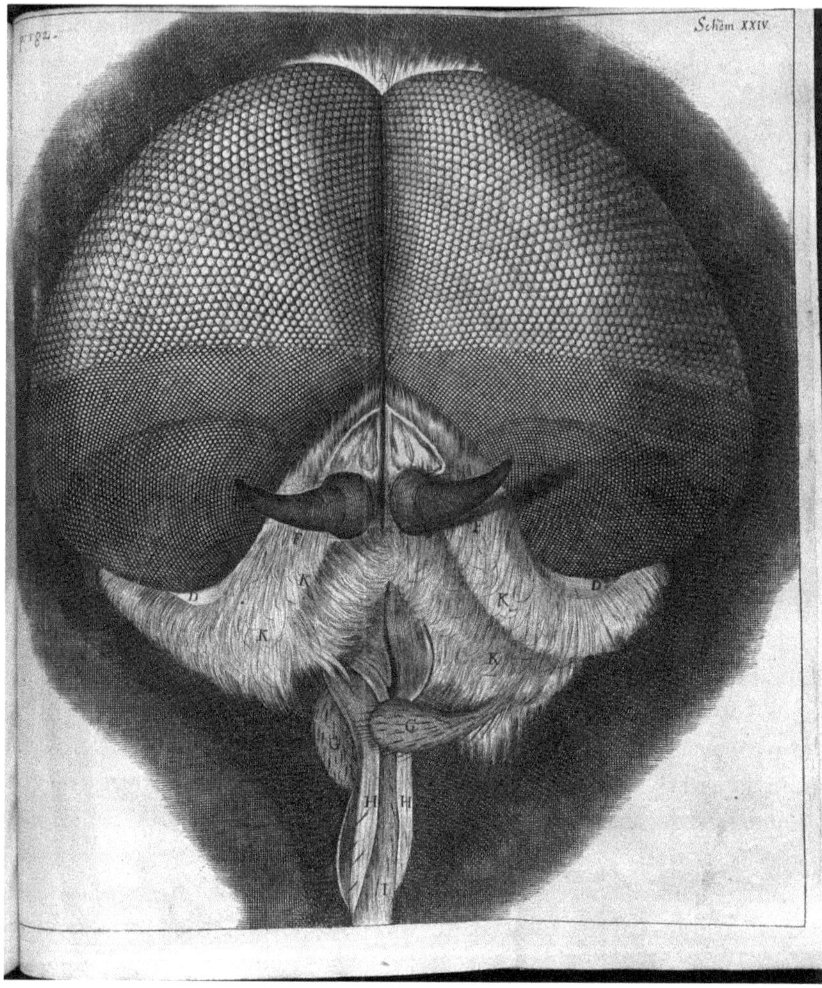

Figure I.7: Anonymous, *Scheme 24*, Engraving, from Robert Hooke, *Micrographia* (London: Printed by J. Martyn and J. Allestry, 1665). By courtesy of the Department of Special Collections, Memorial Library, University of Wisconsin-Madison, Gift of Daniel and Eleanor Albert, Oversize RE26 O62 H66 M53 1665.

towards a delicate bay flower. Stretched along the lower edge of the center of the scroll are the bee's two back legs showing the delicate hairs on the legs and the articulation of the different parts of the legs. The right side of the banner shows a further close-up of the parts of the bee's mouth, a magnified eye, an enlarged stinger, and an antenna that shows its segmented nature, which cannot be seen in the images of the whole bees. In contrast with the stark, up-close view that Hooke presented, the Lincei's magnified insects inhabit a rich allegorical world. These bees were not simply objects to be viewed

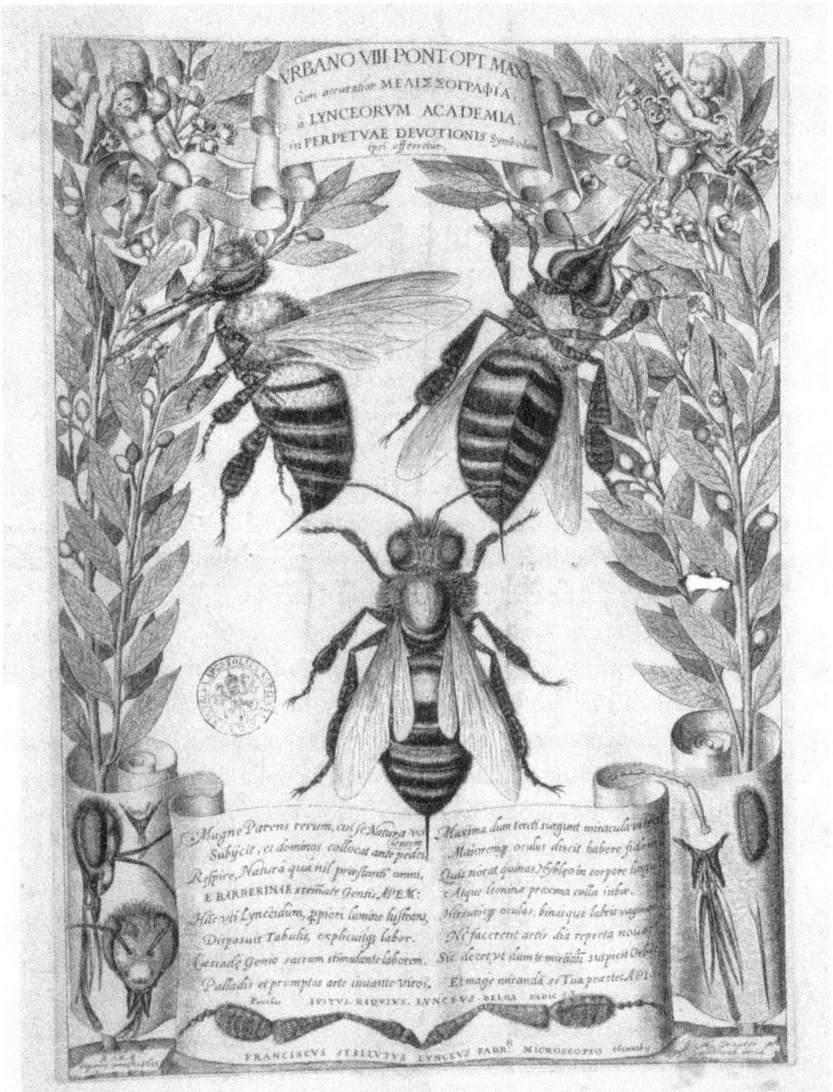

Figure I.8: Francesco Stelluti, *Melissographia*, Engraving, Rome, 1625. The Vatican Library, Vat. lat.9685 – f.117r.

"by varying the degrees of light."[94] Instead, they were part of a larger bid for patronage by Federico Cesi and the other members of the Accademia dei Lincei. Whereas Hooke provided images to aid new users of microscopes, the Lincei used their microscopes to bring glory to their patron. Hooke's production

94 Hooke, *Micrographia*, 175.

of accuracy is in part based on the absence of allegory and performance and instead his insistence on showing what was "plain and evident."

At stake in this discussion is a deeper understanding first of how early modern natural philosophers worked to standardize the production and consumption of knowledge and second of how an artistic medium could foster the communication of new research within a community and produce images that had the visual signs of accuracy.[95] If we look more closely at the production of printed images including a nuanced view of sources, image-text-object relations, and the formal methods used to engrave and print these illustrations, then we gain a deeper understanding of the importance of print in the development of modern culture, and more specifically modern science.[96] As I have begun to suggest, this book does not argue for a sudden revolutionary change in the way that science was practiced; rather, I assert that studying the development of concepts of accuracy in terms of printed images further refines our understanding of how scientific inquiry changed during the seventeenth century because I track the several fronts along which the conventions of the engraved scientific illustration developed. The major concepts under consideration in the following chapters include: the creation of the visual effects of accuracy through careful actions; the development of visual judgment and connoisseurship in this context; the role of a network of individuals in the production of knowledge; the balancing of readers' expectations with representational conventions; the effects of acts of collecting on the creation and circulation of knowledge; and most importantly, the complex relationships among objects, drawings, and printed images, which are neither linear nor straightforward.

The first chapter, "Innocent Witch-craft of Lights," examines the precursors to the production of scientific illustrations by studying books that purported to teach the arts and drawing and engraving to examine how the eye and the hand of the naturalist were trained to consume and produce images that were useful to the production of natural knowledge. These books are used to gain an understanding of how a natural philosopher could accept a two-dimensional representation, often made by someone else, as a useful surrogate for an individual's lived experience of the three-dimensional world.

95 This portion of my argument builds on work on the power images had in the communication of scientific knowledge. See, for example, the essays in: Baigrie, ed., *Picturing Knowledge*; Jones and Galison, eds., *Picturing Science*; and Lefèvre et al., eds., *Power of Images*.

96 Some of the major voices in this discussion are: Eisenstein, *Printing Press as an Agent of Change*; Johns, *Nature of the Book*; and Chartier, ed., *Culture of Print*. For a more recent discussion that takes into consideration the nuanced movement of knowledge through manuscript and print, see: Yale, *Sociable Knowledge*.

The next three chapters offer case studies that examine how accuracy could be recognized and, in addition, study the critical role engraving played in presenting mediated knowledge to a wider audience. By mediated, I mean that the knowledge presented in the image had already been filtered in multiple ways before being translated into a set of engraved lines. The images studied in these chapters were presented as records of what the naturalists and natural philosophers saw and how they saw. The men involved with the production of the works under consideration were witnesses to actual events and saw specific objects. They were not disembodied, objective eyes, but rather embodied, human "I"s. These chapters show how the engraved line was both a precursor to and the medium for the production of knowledge. Each of these chapters examines a different type of mediation: the lens of the microscope and the training of an artist; the collecting practices of early modern natural historians; and the collaborative processes involved in producing a journal. These different forms of mediation required three distinct regimes of accuracy. One regime, based on portrait engraving, produced a visual effect of accuracy in Hooke's *Micrographia*. In a second regime, accuracy in Willughby's *Ornithology* was dependent on visual references to previous images. Finally, in the *Philosophical Transactions*, the effect of accuracy developed through a third regime: the accumulation of images among a community of researchers. Together these four chapters present a more nuanced understanding of the visual communication of science in early modern England.

Chapter two, "Discovering the 'True Form'," uses Robert Hooke's *Micrographia* to examine the intersection of visual conventions for portraiture with the viewing of the microscopic world. In the "Preface" to his *Micrographia*, Robert Hooke asserted that he had discovered "a new visible World" through the help of newly invented optical devices that add "*artificial Organs* to the *natural*." Hooke was also aided by the visual vocabulary developed by engravers for translating a three-dimensional world into a two-dimensional representation of it, and that his awareness of these conventions is what set his illustrations apart from his predecessors.

The third chapter, "Nearly Resembling the Live Birds," looks at the plates in Francis Willughby's *Ornithology* and unearths the sources used by the engravers to produce plates that nearly resembled the life, to use John Ray's phrase. While Ray asserted he was not repeating textual error put forth by his predecessors, such as Gessner and Aldrovandi, he used their illustrations as the basis for his own as well as drawings of live and dead birds that he collected. By uncovering the original sources for the illustrations in Willughby's *Ornithology*, this chapter argues that greater value was placed on recognizable, printed images than on drawings collected by the author.

This conclusion leads to a larger argument about the perceived truth-content of printed natural historical images in the seventeenth century.

The final chapter, "These Rude Collections," consists of a close study of the illustrated articles in the early years of *Philosophical Transactions*. Studying these engravings along with their sources reveals the importance of images to the process of communication within the early modern scientific community and highlights the important role of Oldenburg's network of correspondents in the production of a corporate record of experiments and observations. The accumulation of images in print and manuscript and their publication in the *Philosophical Transactions* created an effect of accuracy for the community of researchers involved in the production and consumption of knowledge about nature.

Taken together these four chapters create a historically grounded picture of the visual traces of accuracy that allowed images to be understood as authoritative and trustworthy. Close examination of images produced for and by the Fellows of the Royal Society reveals the visual and textual markers of accuracy. By combining this qualitative work of humanistic inquiry with more quantitative and macroscopic methods of surveying source materials, this book situates books and periodicals in a larger visual context in order to understand the deep connections between art and science in the period. This combination of different approaches provides readers with an awareness and appreciation of the intricacies of the complex interrelationships between material objects, drawings, engravings, and people that yielded richly illustrated books while also providing a deeper understanding of what accuracy meant in the period and of how accuracy was a way for images to achieve authoritative status.

Bibliography

I. Manuscript Sources

London, Archives of the Royal Society

CLP/2. Classified Papers. Surveying, Opticks, Perspective, Sculpture, Painting, Music, Acoustics Mechanicks.

London, British Library

Add. MS 5260. Drawings of Animals & some parts of Men for comparative Anatomy by Dr.

Sam.l Collins & Dr Edward Tyson. Bought of Dr Collins's Daughter.

II. Printed Primary Sources

"An Account of Some Books." *Philosophical Transactions* 3, no. 39 (21 September 1668): 779–788.

Birch, Thomas. *The History of the Royal Society of London*. London: Printed for A. Millar, 1756.

Boyle, Robert. *A Continuation of New Experiments, Physico-Mechanical, Touching the Spring and Weight of the Air, and Their Effects*. London: Printed by H. Hall for R. Davis, 1669.

Brown, John. "A Remarkable Account of a Liver, Appearing Glandulous to the Eye." *Philosophical Transactions of the Royal Society* 15, no. 178 (1685): 1266–1268.

Casseri, Gulio. *Pentaestheseion, hoc est De qvinqve sensibvs liber*. Venice: Apud Nicolaum Misserinum, 1609.

Collins, Samuel. *A Systeme of Anatomy, Treating of the Body of Man, Beasts, Birds, Fish, Insects, and Plants*. London: Printed by Thomas Newcomb, 1685.

Evelyn, John. *An Idea of the Perfection of Painting*. London: Printed for Henry Herringman, 1668.

—. "A Letter...Concerning the Spanish Sembrador or New Engin for Ploughing, and Equal Sowing All Sorts of Grain, and Harrowing, at Once." *Philosophical Transactions* 5, no. 60 (20 June 1670): 1055–1065.

—. *A Parallel of the Antient Architecture with the Modern*. London: Printed by Tho. Roycroft for John Place, 1664.

—. *Sculptura: Or the History, and Art of Chalcography and Engraving in Copper*. London: Printed by J. C. for G. Beedle & T. Collins, 1662.

Faithorne, William. *The Art of Graveing, and Etching, Wherein Is Exprest the True Way of Graveing in Copper*. London: Published by the Author, 1662.

Haydocke, Richard. *A Tracte Containing the Artes of Curious Paintinge, Carvinge & Buildinge*. Oxford: Joseph Barnes for R.H., 1598.

Hooke, Robert. *Micrographia, or, Some Physiological Descriptions of Minute Bodies Made*. London: Printed by J. Martyn and J. Allestry, 1665.

Pepys, Samuel. *The Diary of Samuel Pepys: A New and Complete Transcription*. Edited by Robert Latham and William Matthews. Berkeley: University of California Press, 1970.

Sprat, Thomas. *The History of the Royal-Society of London for the Improving of Natural Knowledge*. London: Printed by T. R. for J. Martyn and J. Allstrey, 1667.

Tyson, Edward. "Vipera Caudi-Sona Americana, or the Anatomy of a Rattle-Snake, Dissected at the Repository of the Royal Society in January 1682/3." *Philosophical Transactions of the Royal Society* 13, no. 144 (1683): 25–58.

Vesalius, Andreas. *De Humani Corporis Fabrica libri septem*. Basel: Ex officina Johannis Oporini, 1543.

III. Secondary Sources

Alexander, David. "Faithorne, Loggan Vandrebanc and White: The Engraved Portrait in Late Seventeenth-Century Britain." In *Printed Images in Early Modern Britain: Essay in Interpretation* edited by Michael Hunter, 297–316. Farham, Surrey: Ashgate, 2010.

Baigrie, Brian S., ed. *Picturing Knowledge: Historical and Philosophical Problems Concerning the Use of Art in Science*. Toronto: University of Toronto Press, 1996.

Bell, C.F. and Rachel Poole. "English Seventeenth-Century Portrait Drawings in Oxford Collections: Part II Miniature Portraiture in Black Lead and Indian Ink." *Walpole Society* 14 (1925–26): 43–80.

Bourdieu, Pierre. *Outline of a Theory of Social Practice*. Translated by Richard Nice. Cambridge: Cambridge University Press, 1977.

Bury, Michael. *The Print in Italy, 1550–1620*. London: British Museum, 2001.

Chartier, Roger, ed. *The Culture of Print: Power and the Uses of Print in Early Modern Europe*. Princeton, N.J.: Princeton University Press, 1989.

Cooper, Thompson. "Collins, Samuel (bap. 1618, d. 1710)." In *Oxford Dictionary of National Biography*. Oxford: Oxford University Press, 2004.

Croft-Murray, Edward, and Paul Hulton. *Catalogue of British Drawings*, Vol. 1, *XVI & XVII Centuries*. London: Trustees of the British Museum, 1960.

Dackerman, Susan. *Prints and the Pursuit of Knowledge in Early Modern Europe*. Cambridge: Harvard Art Museum; New Haven: Yale University Press, 2011.

Daston, Lorraine. "Epistemic Images." In *Vision and Its Instruments* edited by Alina Payne, 13–35. University Park: Pennsylvania State University Press, 2015.

Daston, Lorraine, and Katharine Park, eds. *Cambridge History of Science*. Vol. 3. *Early Modern Science*. Cambridge: Cambridge University Press, 2006.

Daston, Lorraine, and Peter Galison. *Objectivity*. New York: Zone Books, 2007.

Eisenstein, Elizabeth L. *The Printing Press as an Agent of Change: Communications and Cultural Transformations in Early Modern Europe*. Cambridge: Cambridge University Press, 1979.

Fournier, Marian. *The Fabric of Life: Microscopy in the Seventeenth Century*. Baltimore: Johns Hopkins University Press, 1996.

Fransen, Sietske. "Antoni van Leeuwenhoek, His images and Draughtsmen." *Perspectives on Science* 27, no. 3 (May–June 2019): 485–544.

Freedberg, David. *The Eye of the Lynx: Galileo, His Friends, and the Beginnings of Modern Natural History*. Chicago: University of Chicago Press, 2002.

Gaskell, Roger. "Printing House and Engraving Shop: A Mysterious Collaboration." *The Book Collector* 53 (2004): 213–251.

Godfrey, R.T. "William Faithorne: Portrait Engraver 1616–1691." *History Today* 25, no. 3 (1975): 208–13.

Griffiths, Antony. "Faithorne, William (c. 1620–1691)." In *Oxford Dictionary of National Biography*. Oxford: Oxford University Press, 2004.

—. *The Print Before Photography: An Introduction of European Printmaking*. London: The British Museum, 2016.

—. *The Print in Stuart Britain, 1603–1689*. London: Published for the Trustees of the British Museum, 1998.

Hall, Marie Boas. *Promoting Experimental Learning: Experiment and the Royal Society 1660–1727*. Cambridge: Cambridge University Press, 1991.

Hanson, Craig Ashley. *The English Virtuoso: Art, Medicine, and Antiquarianism in the Age of Empiricism*. Chicago: University of Chicago Press, 2009.

Hanß, Stefan. "The Fetish of Accuracy: Perspectives on Early Modern Time(s)." *Past and Present* no 243 (May 2019): 267–283.

Henderson, Felicity. "Robert Hooke and the Visual World of the Early Royal Society." *Perspectives on Science* 27, no. 3 (May–June 2019): 395–434.

Houghton, Walter E., Jr. "The History of Trades: Its Relation to Seventeenth-Century Thought: As Seen in Bacon, Petty, Evelyn, and Boyle." *Journal of the History of Ideas* 2, no. 1 (January 1941): 33–60.

Hunter, Matthew C. *Wicked Intelligence: Visual Art and the Science of Experiment in Restoration London*. Chicago: University of Chicago Press, 2013.

Hunter, Michael Cyril William. *Establishing the New Science: The Experience of the Early Royal Society*. Woodbridge, Suffolk: Boydell Press, 1989.

Ivins, Jr., William M. *Prints and Visual Communication*. Cambridge, MA: The MIT Press, 1953.

Johns, Adrian. *The Nature of the Book: Print and Knowledge in the Making*. Chicago: University of Chicago Press, 1998.

Jones, Caroline A., and Peter Galison, eds. *Picturing Science, Producing Art*. New York: Routledge, 1998.

Kusukawa, Sachiko. "*Ad vivum* Images and Knowledge of Nature in Early Modern Europe." In *Ad vivum? Visual Materials and the Vocabulary of Life-Likeness in Europe before 1800* edited by Thomas Balfe, Joanna Woodall and Claus Zittel, 89–121. Leiden: Brill, 2019.

—. "The Early Royal Society and Visual Culture." *Perspectives on Science* 27, no. 3 (May–June 2019): 350–394.

—. "The *Historia Piscium* (1686)." *Notes and Records of the Royal Society* 54, no. 2 (2000): 179–197.

—. *Picturing the book of nature: image, text, and argument in sixteenth-century human anatomy and medical botany*. Chicago: University of Chicago Press, 2012.

Latham, Robert. "Forward by the General Editor." In *Catalogue of the Pepys Library at Magdalene College Cambridge*, compiled by Eric Chamberlain. Vol. 3, "Prints and Drawings," Part 2, "Portraits." Woodbridge, UK: D.S. Brewer, 1994.

Lefèvre, Wolfgang et al., eds. *The Power of Images in Early Modern Science*. Basel: Birkhäuser Verlag, 2003.

Lincoln, Evelyn. *The Invention of the Italian Renaissance Printmaker*. New Haven: Yale University Press, 2000.

MacLeod, Catharine, and Julia Marciari Alexander. *Painted Ladies: Women at the Court of Charles II*. London: National Portrait Gallery, 2001.

Marr, Alexander. "Knowing Images." *Renaissance Quarterly* 69 (2016): 1000–1013.

Marr, Alexander, et al. *Logodaedalus: Word Histories of Ingenuity in Early Modern Europe*. Pittsburgh: University of Pittsburgh Press, 2018.

Massey, Lyle. *Picturing Space, Displacing Bodies: Anamorphosis in Early Modern Theories of Perspective*. University Park: Pennsylvania State University Press, 2007.

Ochs, Kathleen H. "The Royal Society of London's History of Trades Programme: An Early Episode in Applied Science." *Notes and Records of the Royal Society of London* 39, no. 2 (April 1985): 129–158.

Oxford English Dictionary, s.v. "accurate," "accuracy," "accurately," and "accurateness." OED Third Edition, December 2011.

Pace, Claire. *Grove Art Online*, s.v. "Disegno e colore." Oxford: Oxford Art Online, 2003. https://doi-org.berea.idm.oclc.org/10.1093/gao/9781884446054.article.T022879. Accessed 18 March 2019.

Park, Katharine. *Secrets of Women: Gender, Generation, and the Origins of Human Dissection*. New York: Zone Books, 2006.

Parshall, Peter. "Graphic Knowledge: Albrecht Dürer and the Imagination." *Art Bulletin* 95, no. 3 (2013): 393–410.

—. "*Imago Contrafacta*: Images and Facts in the Northern Renaissance." *Art History* 16, no. 4 (1993): 554–579.

Payne, Alina, ed. *Vision and Its Instruments*. University Park: Pennsylvania State University Press, 2015.

Poovey, Mary. *A History of the Modern Fact: Problems of Knowledge in the Sciences of Wealth and Society*. Chicago: University of Chicago Press, 1998.

Purver, Margery. *The Royal Society: Concept and Creation*. Cambridge: M.I.T. Press, 1967.

Reinhart, Katherine M. "Richard Waller and the Fusion of Visual and Scientific Practice in the Early Royal Society." *Perspectives on Science* 27, no. 3 (May–June 2019): 435–484.

Richardson, Ruth. *Death, Dissection, and the Destitute*. 2d ed. Chicago: University of Chicago Press, 2000.

Roberts, Lissa, Simon Schaffer, and Peter Dear, eds. *The Mindful Hand: Inquiry and Invention from the Late Renaissance to Early Industrialisation*. Amsterdam: KNAW, 2007.

Shapin, Steven. *A Social History of Truth: Civility and Science in Seventeenth-Century England*. Chicago: University of Chicago Press, 1994.

Shapin, Steven, and Simon Schaffer. *Leviathan and the Air-Pump: Hobbes, Boyle, and the Experimental Life*. Princeton, N.J.: Princeton University Press, 1985.

Shapiro, Barbara J. *A Culture of Fact: England, 1550–1720*. Ithaca, N.Y.: Cornell University Press, 2000.

Sloan, Kim. "Sir Hans Sloane's Pictures: The Science of Connoisseurship or the Art of Collecting?" *Huntington Library Quarterly* 78, no. 2 (2015): 381–415.

Smith, Pamela H. "Art, Science, and Visual Culture in Early Modern Europe," *Isis* 97 (2006): 83–100.

—. *The Body of the Artisan: Art and Experience in the Scientific Revolution*. Chicago: University of Chicago Press, 2004.

Stinjman, Ad. *Engraving and Etching, 1400–2000: A History of the Development of Manual Intaglio Printmaking Processes*. Leiden: Brill, 2012.

Swan, Claudia. "*Ad Vivum, Naer Het Leven,* from the Life: Defining a Mode of Representation." *Word and Image* 11, no. 4 (1995): 353–372.

Walpole, Horace. *A Catalogue of Engravers, Who Have Been Born, or Resided in England*. 2d ed. London: Printed for J. Dodsley, 1786.

Wilson, Catherine. *The Invisible World: Early Modern Philosophy and the Invention of the Microscope*. Princeton, N.J.: Princeton University Press, 1995.

Yale, Elizabeth. *Sociable Knowledge: Natural Knowledge and the Nation in Early Modern Britain*. Philadelphia: University of Pennsylvania Press, 2016.

1. "Innocent Witch-craft of Lights": Developing Visual Judgment through Printed Books

Abstract
This chapter examines the precursors to the production of scientific illustrations by studying books that purported to teach the arts of drawing and engraving to examine how the eye and the hand of the naturalist were trained to consume and produce images that were useful to the production of natural knowledge. My explorations are focused on three texts: *A Book of Drawing* (1647); John Evelyn's *Sculptura* (1662); and William Faithorne's *The Art of Graveing and Etching* (1662). These books are used to gain an understanding of how a natural philosopher could accept a two-dimensional representation, often made by someone else, as a useful surrogate for an individual's lived experience of the three-dimensional world.

Keywords: Drawing, Engraving, John Evelyn, William Faithorne

Fellows of the Royal Society were concerned with how to ensure the accuracy of their images. This chapter investigates the ways in which the effects of accuracy derived from training in how to produce and consume visual materials, particularly drawings and engravings. By making explicit the ways in which artist's manuals and practices informed the construction of early modern visual knowledge of nature, I also aim to show the new ways that authors exerted control over the production of the images associated with their research, which was viewed as an ideal toward which scholars in the seventeenth century strove. The modes of accuracy the Fellows developed as a result of a prolonged engagement with artistic practices and habits of mind grew from several forms of control, including the training and skill required to exert physical control over image production, the intellectual

Doherty, M.C., *Engraving Accuracy in Early Modern England: Visual Communication at the Early Royal Society*. Taylor & Francis Group, 2022.
DOI 10.5117/9789463721066_CH01

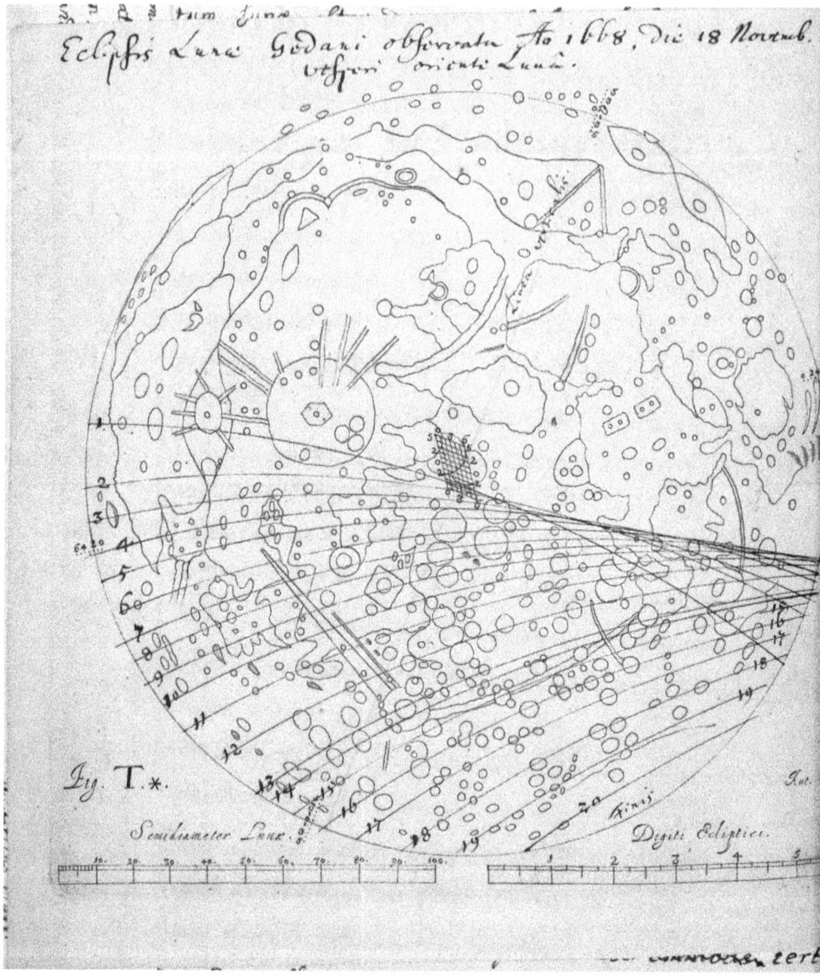

Figure 1.1: Johannes Hevelius, *Lunar Eclipse observed 20 November 1668*, Ink over engraving, Gdansk, 1668. © The Royal Society, LBO/2/128 (326a).

control involved in deploying and refining an emergent visual vocabulary for the circulation of natural knowledge, and the shared controls involved in judging other scholars' visual work as further contributions to a common body of knowledge.

In his book on the history of engraving, *Sculptura: or the History, and Art of Chalcography and Engraving in Copper*, John Evelyn held up Johannes Hevelius (1611–1687) as the model of a scholar who created his own images.

> The learned *Hevelius* has shewed his admirable dexterity in this Art [engraving], by the several *Phases* and other *Ichonisms* which adorn

> his *Selenographi*, and is therefore one of the noblest instances of the extraordinary use of this Talent, for men of Letters, and that would be accurate in the *Diagramms* which they publish in their works.[1]

In Hevelius's case, the accuracy of his images was understood by Evelyn to stem in large part from his skill as a draftsman *and* an engraver.[2] Hevelius signed the plates in *Selenographia* (Gdansk, 1647), "autor sculpsit" indicating that he had engraved the plates himself. Although far removed from the meetings of the Royal Society in London, by the 1660s Hevelius's working habits in Gdansk were apparent to Fellows of the Royal Society in large part through his correspondence with Henry Oldenburg, the Society's Secretary. To send his astronomical observations to Oldenburg, Hevelius used a combination of his skills as an engraver and draftsman. (Fig. 1.1) For instance, using a plate that appeared in *Selenographia*, Hevelius tracked the progress of the Earth's umbra (shadow) across his own map of the surface of the moon.[3] Hevelius's map of the surface of the moon was constructed through repeated viewings. He translated the three-dimensional surface of the moon he saw and transcribed onto a very flat representation of it. This engraving then served as a backdrop upon which he could record his observations of different astronomical phenomena. Whereas in 1647 the map itself was the conveyor of new information, by the 1660s it was instead the medium for his reports of new observations. Hevelius's hybrid print-drawing was both a record of his intellectual work and his manual work. Furthermore, it quite literally visualized the work of both his eye and his hand for his correspondents. Evelyn rightly singled Hevelius out as "extraordinary...for men of Letters" (83). The questions this chapter asks consequently hinge on the desire to the master the means of production in order to create the effects of accuracy. First, then, how did English natural philosophers learn to draw in the middle of the seventeenth century? In addition, what did they know about the engraving process? And finally, how did the combination of skill and craft knowledge shape

1 Evelyn, *Sculptura*, 82–83. Subsequent citations to this book will be made parenthetically in the text.

2 Evelyn read Hevelius's *Selenographia* shortly after it was published in 1647 and made notes on it in his commonplace book. Hunter, "John Evelyn," 74.

3 This engraving with manuscript additions was sent to Henry Oldenburg along with textual and tabular descriptions of his observations of a lunar eclipse on 18 November 1668 (N.S.). The letter and table are preserved in the Early Letters series in the Archives of the Royal Society (EL/H2/17). The image is housed with the fair hand copy of the letter (LBO/2/326). All three parts of the report are reproduced in: *Correspondence of Henry Oldenburg*, vol. 5, 142–148.

their judgments about what images fit a model of accuracy based in that practical episteme?

While access to the craft knowledge and trade secrets bound up in the process of making and producing an engraving had not always been readily available, an explosion of printed manuals in the middle of the seventeenth century provided an increasingly broad audience with detailed instructions in the arts of drawing, painting, and engraving. Mastery of the arts of drawing and engraving was no longer limited to the confines of workshops and academies.[4] Instead, a model of self-education through print developed. As Kim Sloan has shown, in the first half of the century drawing was primarily a pursuit of "royalty and those closest to them."[5] Beginning during the Civil War and increasingly after the Restoration, this changed. Fellows of the Royal Society were reading and producing books on the history and practice of engraving and the practice of drawing. Their interest in these texts grew out of exposure to Continental art manuals, in some cases during time spent on the Continent in exile during the Civil War.[6] Early examples of illustrated books on drawing include: Albrecht Dürer's *Underweysung der Messung* (Nuremberg, 1525); a series of prints published by the Carracci brothers in Rome before 1616; Jean Cousin's *Livre de Pourtraittre* (Paris, 1608); and *Diagraphia, sive Ars Delineatoria* published by Joannes Janssonius in Amsterdam (1616).[7] The first treatise dedicated solely to the arts of etching and engraving by Abraham Bosse was published in in Paris in 1645.[8] English printmakers and booksellers, some of whom were similarly on the Continent during the Civil War, translated and reprinted these early examples.[9]

My explorations of the explosion of printed books on the history and practice of art in England are focused on three types of texts: first, those that purport to teach drawing; second, those that recount the history of a trade; and finally, those that recount the practice of a trade. The titles I explore

4 For discussions of traditional art education, see: Goldstein, *Teaching Art* and Flick, *Masters & Pupils*.

5 Sloan, *'A Noble Art,'* 11. She elaborates on this idea in chapter one of the book, "Virtue, *Virtuosi* and Views."

6 This was certainly the case for Evelyn. Hunter, "John Evelyn," 67–98

7 In her catalogue raisonné of the prints by the Carracci family, Diane DeGrazia Bohlin states that the *Scuola perfetta per imparare a disegnare* was never formally published. Bohlin, *Prints and Related Drawings*, 410. The date given in most library catalogues for the publication of the *Scuola perfetta* is ca. 1620. However, given the inclusion of plates from this book in *Diagraphia*, which was published in 1616, this date is too late.

8 Hults, *The Print*, 212 and 285.

9 As will be discussed in detail later in this chapter, William Faithorne was one such printmaker who spent time in Paris in exile.

below exemplify three categories of books that Fellows of the Royal Society were reading (and writing) in the second half of the seventeenth century. My exemplars are *A Book of Drawing, Limning, Washing and Colouring of Mapps or Prints* (1647), John Evelyn's *Sculptura: or the History, and Art of Chalcography and Engraving in Copper* (1662), and William Faithorne's *The Art of Graveing and Etching Wherein is Exprest the True Way of Graveing in Copper* (1662). Evelyn's publication of *Sculptura* and his (unpublished) translation of Bosse grew out of the Royal Society's pursuit of Francis Bacon's call for a history of trades. Although seemingly aimed at different audiences ("Ingenuous Gentlemen and Youths,"[10] "such as are addicted to the more *Noble Mathematical Sciences*,"[11] and "all Lovers of Arts"[12]), all three of these books were read by Fellows of the Royal Society.[13] Taken together these three books provide a historical and pragmatic justification for the study of drawing and engraving, an aesthetic education which allowed for the development of visual judgment, and practical guides for learning these skills. All three texts are simultaneously concerned with the production and consumption of images.

These texts are tied together by their emphasis on the importance of good judgment to the successful creation of images. This chapter explores how practical manuals as well as historical tracts on the arts of drawing and engraving helped readers develop their skills as draftsmen and engravers as well as their ability to judge the quality of their own and others' work. Throughout these texts, judgment is stressed as a critical component to the successful production *and* consumption of images. Being a good judge of the quality of images was a necessary precursor to being able to produce images as it was essential to first choose good models to follow. Drawing manuals such as *A Book of Drawing* provided the young draftsman with ready models. *Sculptura* provided a detailed history of printmaking that noted key works of each artist mentioned, and thus led the reader to develop a vocabulary as well as an aesthetic sense to shape his own appreciation of high-quality printed works. Faithorne's directions to would-be printmakers continually stressed the necessity of discernment and informed decision-making throughout the process of preparing to make a print and in so doing looked forward to the circulation of published images. All three books emphasize the necessity

10 *Book of Drawing*, title page.

11 Evelyn, *Sculptura*, A2v. Emphasis in the original.

12 Faithorne, *Art of Graveing*, A2v.

13 As will be discussed later in this section, Evelyn's text was dedicated to Robert Boyle, another founding Fellow of the Royal Society. For Robert Hooke's ownership of a copy of *A Book of Drawing* and for his connections to William Faithorne, see chapter two of this book.

of first training the eye to know what you are looking at and to know how to select good models to follow—that is, as a seventeenth-century natural philosopher, your gaze must be trained to consume images before your hand is trained to produce them.

Two major changes occurred in this time and place that affected how images were consumed and what it meant for an image to be useful for the circulation of knowledge. These changes were related to authors and audiences. First, the authors of texts about images expanded beyond practicing artists to include connoisseurs and historians. Fellows of the Royal Society, like John Evelyn, began writing books on the arts of drawing and engraving.[14] These books were no longer being written by practicing artists like Nicholas Hilliard, who wrote his *Treatise Concerning the Arte of Limning* in 1598 or 1599.[15] Further, authors were no longer producing all-in-one volumes that purported to unlock the secrets of all the arts which hinted at revealing methods but provided very little in the way of concrete instruction, such as John Bate's *The Mysteryes of Nature and Art* (London, 1634), which included engraving among other mysteries such as fireworks and fountains.[16] Works like *A Book of Drawing* and Faithorne's *Art of Graveing* provided detailed, step-by-step instructions beginning with the necessary tools through the whole image making process. Unlike previous drawing manuals, like those by Jusepe Ribera (1591–1652), Guercino (1591–1666), Stefano della Bella (1610–1664), Abraham Bloemaert (1566–1651), and Peter Paul Rubens (1577–1640) as discussed by Caroline Fowler, which contained only images, these books combined text and image to provide a more detailed instruction for a non-specialist audience.[17] Second, the audience for information about these arts broadened dramatically with these publications. The appearance of printed books on drawing and engraving took knowledge that had previously been transmitted orally in the case of the apprenticeship model and in manuscript form in the case of courtly education in drawing and made it available to a larger public.

14 Sachiko Kusukawa has also discussed Evelyn and his interest in prints and drawings as it relates to the early Royal Society's attitudes toward images. Kusukawa, "Early Royal Society," 360–365.

15 Bermingham, *Learning to Draw*, 19–26.

16 The full title of Bate's work gives a sense of the range of topics: John Bate, *The Mysteryes of Nature and Art: Conteined in four severall Tretises, the first of Water Works, the second of Fyre Workes, the third of Drawing, Colouring, Painting, and Engraving, the fourth of divers Experiments as wel serviceable as delightful: partly collected and partly of the Authors Peculiar Practise, and Invention.*

17 Fowler, *Drawing and the Senses.*

Even with the beginnings of formal art academies, instruction in how to draw was still reserved for a group of insiders.[18] As we saw with Evelyn's discussion of Hevelius, one corollary development was that image-making skills came to be seen as necessary for the natural philosopher and natural historian.

Although the texts considered in this chapter claimed to be about making images, they were as much about how to consume them. The skills of the connoisseur had to be developed alongside those of the draftsman. The self-education model supported by these books provided initial exemplars, but also sought to enhance readers' ability to choose good models to follow. Consequently, visual and manual expertise were mutually reinforcing as readers were presented with good models that helped them improve their own skills and gave them materials against which to judge their own work. In addition to using manuals to learn how to make images, readers were learning to consume them. The production and consumption of images increasingly became imbricated processes as each required the development and deployment of visual judgement.

The emphasis on visual education in these books focused on prints, so the system of translating three dimensions into two was necessarily based on using lines. William MacGregor has written about the important role of prints in the education of young artists in early eighteenth-century France.[19] The observations he made about the place of prints in visual education was well established in the seventeenth century, as can be attested by the burgeoning trade in printed drawing manuals aimed at non-specialist audiences. In addition, Fellows of the Royal Society, such as Robert Hooke, stressed the importance of copying prints in learning to draw.[20] The engraved line then served both as the starting point of learning to make images for research and for natural philosophers as the end point in printed books. Similarly, the intellectual process of developing criteria by which to judge the quality of images was positioned as a preliminary to the manual process of creating images, both drawings and prints.

The books under consideration also provide insight into how artists constructed an understanding of their private practice for public

18 For instance, see Flick's discussion of the Carraccis' academy: Flick, *Masters & Pupils*, 99–125. Karen-edis Barzman discusses the Accademia del Disegno's connections to the Medici court and Florentine humanism more broadly. Barzman, *The Florentine Academy*, esp. chap. 5, "*Disegno* as a Disciplinary Practice: The Academy School," 143–180.

19 MacGregor, "The Authority of Prints," 389–420.

20 Felicity Henderson discusses how Hooke "set his assistant Henry (Harry) Hunt the task of copying images, presumably acting as drawing tutor himself." Henderson, "Robert Hooke," 408.

consumption through print. Texts such as William Faithorne's *The Art of Graveing* presented readers with a clear view into a space where artists purportedly worked to control the actions of their hands and instruments. This move towards openness finds parallels in the Royal Society's history of trades project in general, and more specifically in the publication of experimental set-ups by Robert Hooke in his *Micrographia* and Robert Boyle in *Medicina Hydrostatica* and *New Experiments Physico-Mechanicall.*[21] These texts all share a visual gesture towards openness that claims to show the reader everything they need to know to replicate the desired results.

Understanding Fellow's exposure to and engagement with the arts of drawing and engraving sets the stage for understanding how they could come to accept a printed image as an accurate representation of an object or phenomenon. Each of these three case studies exposed different aspects of the intricate interconnections between drawing and print. By bringing readers into the artist's studio, *A Book of Drawing* provided instruction that seemed to mimic that of traditional workshop practices, which in turn allowed readers to believe their drawings participated in a long artistic legacy. Through detailed discussion of individual engraver's works Evelyn produced a history of engraving that was based on the Baconian model of how to write natural history. His collection of facts and instances allowed his readers to develop their own visual judgment to aid them in the consumption and production of images. *The Art of Graveing* worked on two levels to produce accuracy: first, it developed readers' abilities to judge errors in their own works and second, it opened a window onto artisanal practices that allowed readers to gain an understanding of how engravers worked to prevent errors from occurring in the first place.

This new model of self-education through the use of printed books meant that no single model for the production of accuracy existed as skills formerly held as trade secrets were put to new ends. In addition to developing manual skills, these books explicitly developed their readers' ability to judge the quality of images, which led to their willingness to accept reproductions as accurate representations for the purpose of communicating knowledge because of the emergence of basic visual conventions to communicate regular characteristics of visual material. As the other chapters in this book explore, these developments resulted in different regimes of accuracy across different fields of inquiry.

21 These are just two examples of Boyle's publications that include images of his equipment and experimental set-ups.

The Magic of Projection

The translation of a colorful, three-dimensional world into a series of two-dimensional black lines represents the first key element in drawing. This translation is neither simple nor straightforward, and the processes involved in accomplishing its effects gave rise to an explosion of books published on topics once reserved for the closed relationship of the master and the apprentice. For Fellows of the Royal Society to effectively generate and consume images that were useful for the production and circulation of knowledge, they first had to train their hands and eyes. These books helped readers to unveil the "the *Magic*, and innocent *Witch-craft* of lights and shades" of good drawing to use John Evelyn's phrase (107). This chapter traces how these manuals opened the doors for Fellows to investigate this magic.

Albrecht Dürer played an important role in Evelyn's narrative about the history and future of engraving in the context of the early Royal Society. Dürer served as a model for the particular combination of talents that Evelyn sought to cultivate in his readers because even from a young age he was "performing such things as might shame most of the best Masters" (36) and was a "Prodigie of science" (63). In the context of this combination, Dürer was praised for the high caliber of his drawings, engravings, and woodcuts as well as his command of and development of the rules of perspective, that is "of giving things their natural distance" (38). Evelyn placed a great deal of importance on this ability to render a sense of three-dimensionality in an image. Being able to represent a three-dimensional object in a two-dimensional rendering was essential for the communication of natural knowledge. Whether that object was a dissected body or a flea, the ability to create a convincing likeness was key.

Much of Evelyn's most pointed criticisms and most effusive praise of artists were linked to their failures and successes in translating a three-dimensional object into a two-dimensional representation of it. When discussing the necessity of learning to draw prior to attempting to make a print, Evelyn stressed the importance of capturing a sense of three-dimensionality:

> The pen is therefore both the first, and best instructive, and has then (as all the other kinds) attain'd its desired end, when it so deceives the eye by the *Magic*, and innocent *Witch-craft* of lights and shades, that elevated, and solid bodies in *Nature*, may seem swelling, and to be embossed in *Plano*, by Art (107).

The wonder of representation for Evelyn then lay in its ability to deceive. It was through learning to use hatching well that his reader could hope to achieve such a deception.[22] Evelyn gave only vague directions as to how to manage this feat conceding that not all of his readers would be able to accomplish it without "the help of the *rule* and *compass*; every man being not an *Apelles* or *Pyrogoteles*" (108). Given the limitations of his audience's skills, Evelyn returned to the suggestion that "the best expedient to gain a mastery in this Address, will be to imitate such *prints*, and *cuts*, as are most celebrated for this perfection" (108). He then provided a list of artists who should be copied, reinforcing the message that the reason to learn about past artists and their work was to allow you to know which examples to follow in order to train your eye prior to training your hand.

While the majority of the chapter on drawing and design worked to convince his readers of the necessity of drawing as a precursor to engraving, Evelyn dealt directly with the problem of projection in the short section that provided concrete instructions for pursuing this art. He began by stating that he could well end the chapter "having abundantly vindicated the necessity and worthiness of *Designe* and *Drawing*, as it is prævious, and introductory to the Art of *Chalcography*" however he was compelled to include a more detailed discussion of "*Hatches* and stroaks":

> ...pretending (at least very ingeniously hinting) to a method, how by a constant, and regular certitude, one may express to the eye, the Sensation of the *Relievo*, or *extancie* of objects, be it by one, or more hatches, cross and counter, we think not impertinent here to recite, as briefly as the demonstration will permit (118–119).

What followed was an illustrated description of how to capture a sense of the three-dimensionality of objects based on the work of Abraham Bosse. The section included one of the two illustrations in the book, this one taken directly from Bosse's treatise on the practice of perspective aimed at printmakers, the other being a sample mezzotint from Prince Rupert's own hand.[23] (Fig. 1.2) Evelyn had met Bosse during his time in France and recorded visiting him to discuss exactly this issue of perspective.[24] The focus here is not on linear perspective and the representation of large

22 In her discussion of this passage, Sachiko Kusukawa highlights how for Evelyn this witchcraft was "something positive." Kusukawa, "Early Royal Society," 362.

23 Bosse, *Moyen Universel*. This plate appears as figure thirty-one.

24 "I went to that excellent Ingraver *Du Bosse*, for his instruction about some difficulties in *Perspective* which he delivered in his booke." *Diary of John Evelyn*, ed. E.S. de Beer, vol. 2, 568 (31 December 1649).

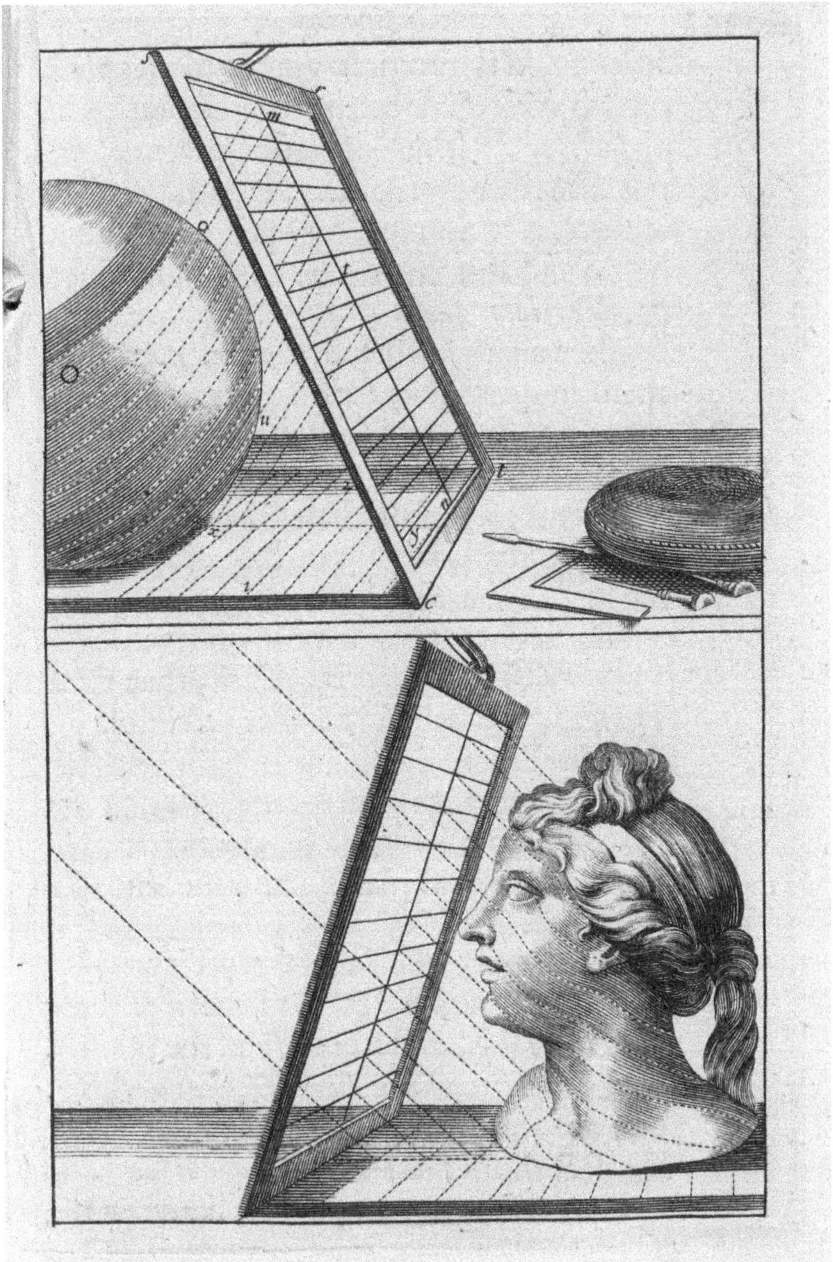

Figure 1.2: Anonymous, *How to draw a sphere and a head*, facing page 120, Engraving, from John Evelyn, *Sculptura* (London: Printed by J. C. for G. Beedle & T. Collins, 1662). The Bodleian Libraries, University of Oxford, 8vo G.119 Linc.

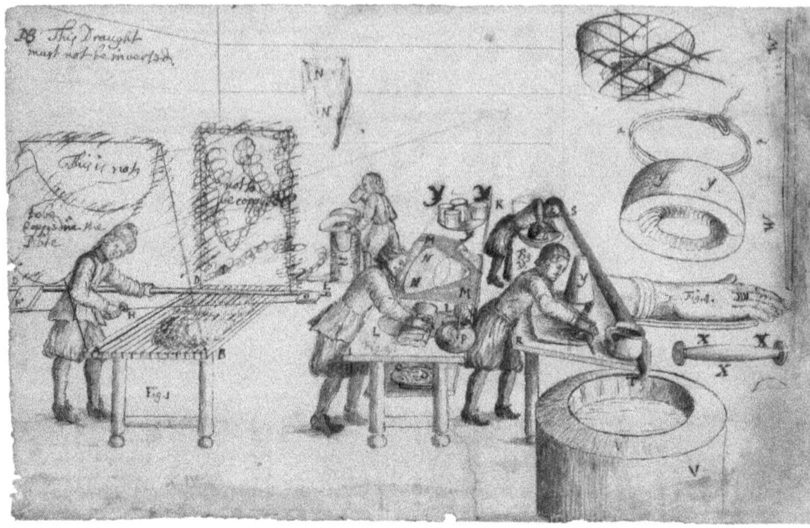

Figure 1.3: Robert Hooke, *Felt-makers at work*, Ink on paper, Undated. © The Royal Society, CLP/20/96.

architectural spaces, but rather on the depiction of objects. As with most instructions on the art of drawing, these directions begin with mastering the representation of an abstract form before moving on to human bodies, in this case a head.

For the natural philosophers Evelyn aimed to reach with his treatise, these technical instructions and the visual depiction of both the method and the instruments would have provided a basis for the translation from three to two dimensions that was central to the conjuring of visual magic Evelyn encouraged them to pursue. These directions would have allowed them to explore the possibilities of creating likenesses that produced the illusion of accuracy, be they geometrical, astronomical, natural historical, or otherwise related to the Fellows' efforts to promote the accumulation and distribution of natural knowledge. While Evelyn praised Dürer's ability to give "things their natural distance," it is not an Albertian sense of spatially focused linear perspective he stressed for his readers. Instead, they were making and consuming images of things. That is, the key information in an image was often tied to the relative size of objects or the relationship of parts to the whole, not the creation of three-dimensional spaces. For example, Robert Hooke's drawing of a felt maker's workshop, seemingly prepared for an engraving although there is not one extant based on it, used some of the basic concepts of one-point perspective, but the lines of the tables do not actually converge

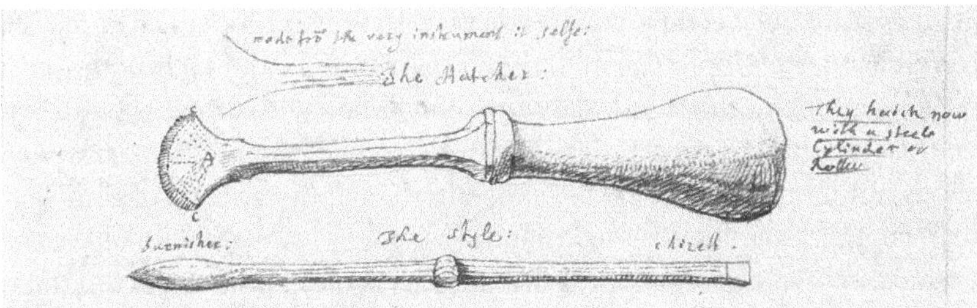

Figure 1.4: John Evelyn, *The Hatcher*, Ink on paper, ca. 1659. © The British Library Board, Add. MS 78340, fol. 154.

on a single point.[25] (Fig. 1.3) Instead his image hinted at a sense of the three-dimensional space, but the focus is on what the people are doing and the tools they are using. The shading on the objects in the drawing is consistent with a single, strong light source coming from the right. While the viewer might not be able to imagine the space in which these men work, he could envision holding these tools or pressing the fabric to remove excess water and achieve the desired consistency of the felt.

For Evelyn, the magic of an image like this was dependent on the artist's use of parallel and crossed lines to give a vivid sense of the three-dimensionality of the objects. The accuracy of an image like Hooke's then is tied up with its ability to convey the information that is most useful for the viewer. In his own work, Evelyn used these same techniques to depict the objects he described. (Fig. 1.4) For example in an unpublished description of Prince Rupert's method for making a mezzotint, Evelyn included a drawing of the hatcher, which is used to work up the plate. Using a mixture of hatching and short pen strokes, Evelyn clearly conveys the roundness of the handle and the thickness of the metal shaft. *Sculptura* worked to develop the viewer's ability to judge the draftsman's or engraver's skill at creating this illusion. In other words, accuracy resides in the combination of the maker's skill at rendering three-dimensional objects and the viewer's skill at recognizing and seeing through the magical work of engraving.

A Young-Mans Time Well Spent

While the previous section focused on the practices tied to the translation of three-dimensional space into two-dimensional rendering with lines on paper, the following section approaches the effects of accuracy through

25 For a discussion of the context in which this drawing was made, see: Henderson, "Robert Hooke," 415–417.

an analysis of the weight attributed to identifying and copying a fit model, as well as to the proper use of drawing instruments and technique. John Evelyn begins the chapter in *Sculptura* concerned with drawing by stressing that "he shall never attain to the excellency of a good *Chalcographer*, who is not more then ordinarily skill'd in the faculty and art of *Drawing*" (103). Furthermore, he stressed that drawing was intimately linked to the production of knowledge calling drawing "a visible expression of the Hand resembling the conception of the mind." The goal of a drawing, for Evelyn, was to translate the intellectual work of the natural philosopher, through manual labor, into something that others would recognize. Drawing is the precursor to printing for both pedagogical and practical reasons. The cost, availability, and technical requirements of printing materials meant (and still means) that artists and natural philosophers would produce drawings (often multiple drafts) before committing their designs to a copper plate or woodblock. However, Evelyn provided very little in the way of practical advice. For that his readers had to turn to the myriad manuals aimed at teaching drawing to those outside of the traditional artistic training model of apprenticeship. I take as my exemplar *A Book of Drawing, Limning, Washing or Colouring of Mapps and Prints*, first published in 1647 and later retitled *Albert Durer Revived*.[26]

A Book of Drawing shows what a drawing manual taught and the visual source material available to young men learning to draw. It stands out from similar books because of its long publication history, extending from the first edition in 1647 at least as far as 1731 (the latest printing I have found).[27] I argue that the longevity of the publication is tied to its connection with Albrecht Dürer, who was positioned as the book's source of inspiration and a model for a method of learning to draw. The specter of Dürer was

26 A version of this material was previously published in Doherty, "*Young-Mans Time*," 51–78. Throughout this chapter I am using the text from the 1652 edition at the Huntington Library, which is available as part of *Early English Books Online* (Wing (2nd ed., 1994) / B3705AB). I have compared this text with the 1647 and the differences are only in the type setting, not in the content. Although Thomas Jenner is often listed as the author of this text in library catalogues, I will not refer to him as the author in this chapter. Because the 1616–1620 edition published by Compton Holland is known only through its frontispiece, it is unclear to me whether the text in the 1647 edition, and all subsequent ones, is Jenner's work, or Holland's, or some other anonymous author's. A copy of the Holland frontispiece is preserved at the Huntington Library in Richard Bull's extra-illustrated copy of *Biographical History of England* (call no. 238000 v.2:85). Two others are in the collection of the British Museum (1933,0614.30 and J,11.152). The first edition to add "Albert Durer Revived" to the title was published in 1679 (Library of Congress, NC730 .B6 1679 (Rosenwald Coll.)). Further quotations from this book will be cited parenthetically in the text.

27 For a complete list of the editions I have consulted see: Doherty, "*Young-Mans Time*," 70–74.

continually invoked in books aimed at spreading information about the history and practice of art.[28] A portrait of Dürer served as the frontispiece to all the editions of the book and the text on the plate praised Dürer as "the verie prime painter and graver of Germany." (Fig. 1.5) Dürer was not the literal source for instructions on how to draw provided in the book; rather, the name "Albert Durer" served as a proper name for a set of practices and a method for learning to draw, which involved studying prints that offered a systematic approach for young men inclined to learn to draw nature. Readers were encouraged to follow the practices common within artisans' workshops with only a surrogate master guiding them. Having "Albert Durer" guide readers through the workshop allowed those without access to that space or the courtly ones of Hillyard a chance to gain an understanding of these practices and apply them to their own pursuits. This apparent access to the workshop practices of Albrecht Dürer gave the book a sense of legitimacy as readers seemed to be gaining the secrets of Dürer's studio. At the same time, access to the workshop provided readers with a claim to accepted practices that in turn lent their own drawings a sense of value.

While this book remained in print the longest, it was by no means the only manual aimed at teaching people to draw outside of the apprenticeship system or courtly circles. Several London drawing masters published manuals as well. For instance, Alexander Browne (d. 1706), a well-regarded London drawing master, whose pupils included among others Samuel Pepys's wife, published a few manuals during the 1660s and 1670s.[29] The illustrations in Browne's manuals were drawn from Italian sources.[30] Similarly, the images in *A Book of Drawing* were not drawn from Dürer's published works but rather from previously published drawing manuals by Jean Cousin (ca. 1500–ca. 1560), Odoardo Fialetti (1573–1638), Joannes Janssonius (1588–1664), and Agostino Carracci (1557–1602).[31] (Fig. 1.6) Readers were encouraged to follow the common practice within artisans' workshops of copying prints with only a surrogate master to guide them.[32] Further manuals published around

28 For more on Dürer's legacy see: Bartrum, *Albrecht Dürer* and Bubenik, *Reframing Albrecht Dürer.*

29 See for instance, Browne, *The Whole Art of Drawing*; Browne, *Ars Pictoria* (a second edition was published in 1675); and Browne, *A Compendious Drawing-Book.* Pepys recorded his wife starting lessons with Browne on May 7, 1665. *Diary of Samuel Pepys*, vol. 6, 98 (7 May 1665).

30 Bermingham, *Learning to Draw*, 47.

31 Doherty, "*Young-Mans Time*," 51–78.

32 For a detailed discussion of the use of prints in the education of Italian Renaissance artists, see Lincoln, *Invention of the Italian Renaissance Printmaker*, esp. chap. 1, "Andrea Mantegna," 17–44.

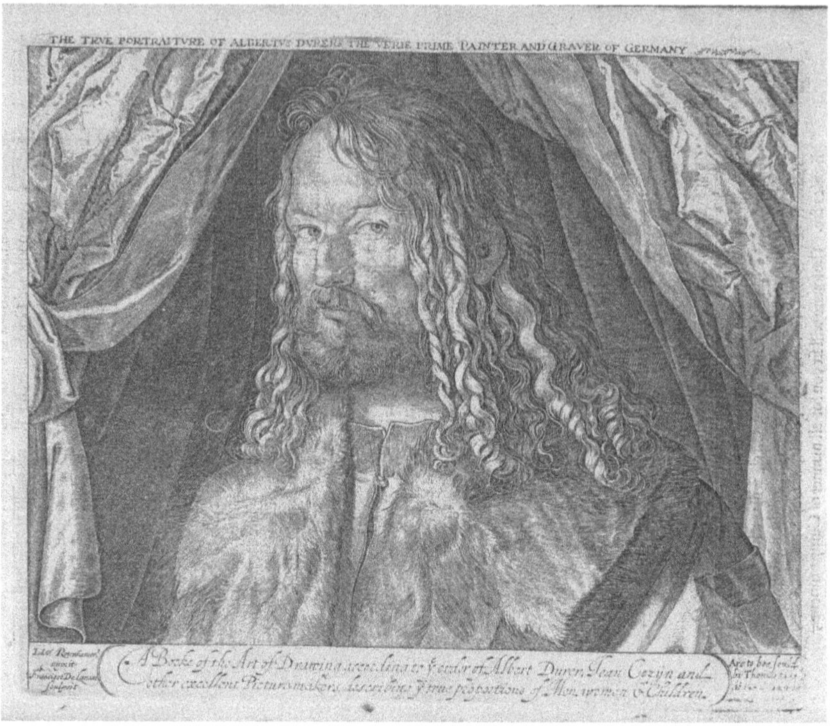

Figure 1.5: Francis Delaram, *Portrait of Albrecht Dürer*, Engraving, from *A book of drawing, limning, washing or colouring of maps and prints, or, The young-man's time well spent* (London, Royal Exchange: Printed by James and Joseph Moxon, for Thomas Jenner, 1647). Yale Center for British Art, Paul Mellon Collection, NC730 .B6 1647+ Oversize.

this same moment and in the same style included: *A Booke of Drawinges. Performed according to the best order for use & Brevity that is yet Extant* (London, 1650); *A Booke of Portraicture* (London, 1665); *Academia Italica* (London, 1666); and *The Excellency of the Pen and Pencil* (London, 1668). What these manuals all had in common was their aim of providing readers with guides for learning to draw "collected out of the choicest Italian and German Authors" as Browne put it on the title page to his first manual published in 1660. Through using these books readers were to develop both their skills as draftsmen and their ability to judge a good image from a bad one.

A Book of Drawing then provided those wishing to learn to draw with a set of well-selected models and a famous surrogate master to help them in their pursuit. The images worked with the text to teach a young gentleman the art of drawing and impressed upon him a way of understanding the translation of the three-dimensional world into two-dimensional representations of it. According to Evelyn's recommendations in *Sculptura*, the key to learning about engraving was to study examples of work by skilled artists: *A Book of*

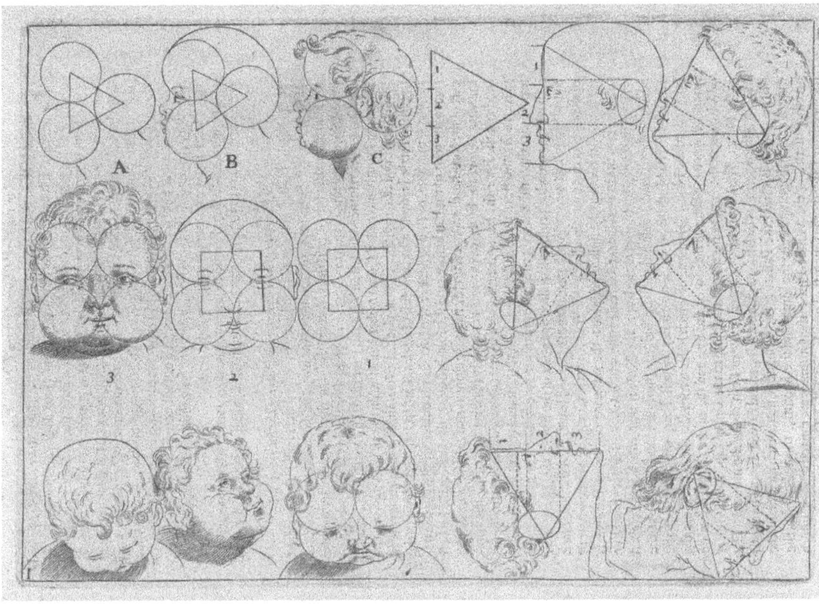

Figure 1.6: Anonymous, *How to draw faces*, Engraving, from *A book of drawing, limning, washing or colouring of maps and prints, or, The young-man's time well spent* (London, Royal Exchange: Printed by James and Joseph Moxon, for Thomas Jenner, 1647). Yale Center for British Art, Paul Mellon Collection, NC730 .B6 1647+ Oversize.

Drawing was based on the same premise.[33] These two books set up a mutually reinforcing relationship between printmaking and drawing. While Evelyn saw the art of drawing as "an absolute requite to that of *Chalcography*," *A Book of Drawing* set up copying from prints as a key to learning to draw. The format of every edition of the book reinforced this methodology as each page of text faces an image.

The title page of the first edition of *A Book of Drawing* (1647) explained for whom the book was intended and what the author considered its usefulness for those learning to draw. The subtitle, *A Young-Mans Time Well Spent*, made it clear that it was an educational tool for young gentlemen and that learning to draw was not just an idle waste of time, but a most worthwhile pursuit. The title page promised the young man "the Ground-work to make him fit for doing any thing by hand, when hee is able to Draw well." It went on to state that the book would be "very usefull for all Handicrafts, and Ingenuous Gentlemen, and Youths," increasing the sense that this book was aimed at curious, young gentlemen of leisure. The title page concluded with

33 For a discussion of copying in the context of the Royal Society and early modern learning, see: Fransen and Reinhart, "Practice of Copying," 214–216.

a moral Latin epigram, "*Infœlix qui pauca sapit, spernitque Doceri*" ("[He is] unfruitful who knows little and refuses to be taught").[34] This epigram emphasized the importance of education for a useful member of society and that learning to draw was a part of an ideal educational process. The title page gave the impression that this book would provide a young man with a useful occupation and prepare him for whatever lay ahead.

Without a drawing master to guide them in person, those learning to draw with a book first had to recognize, and then procure and learn to handle the appropriate tools. Whereas Henry Peacham (1546–1634) in his *The Art of Drawing with the Pen* (London, 1607) began with an introduction to the history of drawing and painting, *A Book of Drawing* began with the practicalities of selecting and using the "necessary Instruments for Drawing" (2). It is not until page ten that Peacham discusses the instruments needed for drawing and then only briefly. In *A Book of Drawing*, on the other hand, the text went on at some length about the different types of tools for making lines as well as the conditions in which each pencil or pen should be used. This book served as a practical guide for its readers and the tools and their uses were not obvious to those outside the realms of traditional art education. Faulty tools would produce faulty results, which were unacceptable, especially for the natural philosopher.

The 19 plates in *A Book of Drawing* provided a variety of well-selected images to be copied. (Fig. 1.7) The author emphasized that only approved masters should be used as models: "...which you must diligently observe, and you shall by little and little finde out, in good Masters workes which you should chiefly desire to imitate, and not botchers" (6). The reader's eye was likewise being trained to recognize high-quality work. Good judgment was to be inculcated by studying illustrations from "good Masters workes." The images used were "patterns" which should be scrutinized closely and copied exactly; the term was used eleven times (in the phrase "according to your pattern" six times).[35] Success could only be achieved "by imitating your patterne exactly" (12), which served as the ultimate guide as to what the final drawing should

34 This translation is from Shuffelton, *Codex Ashmole 61*, 71.

35 From *A Book of Drawing*: "...lightly draw the out-stroak, or circumference of the face just according to the bignes of your pattern, making it to stand fore-right, or to turn upwards or downwards according to your patterne..." (6); "...but onely make it at first but a bare circumference, turning this way, or that way, according as the pattern doth..." (6); "...always make your curle to bend, and turne exactly according to the patterne..." (6); "...then if you have done that right, part the fingers asunder, or close, according to your pattern, with the like faint stroake..." (8); and "last of all draw the armes, and then the hands, either joyned to the body or separated from the body, according to your patterne" (10).

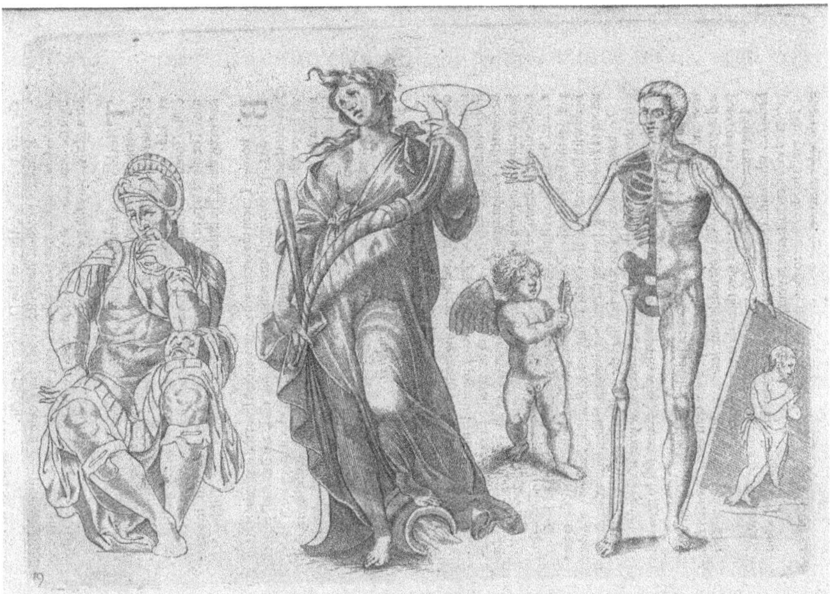

Figure 1.7: Anonymous, *A collection of figures*, Engraving, from *A book of drawing, limning, washing or colouring of maps and prints, or, The young-man's time well spent* (London, Royal Exchange: Printed by James and Joseph Moxon, for Thomas Jenner, 1647). Yale Center for British Art, Paul Mellon Collection, NC730 .B6 1647+ Oversize.

look like; the final drawing thus should "resemble the patterne you draw it from." "Your pattern" should be consulted to better understand how to show the mass of the body through covering draperies (14). In these instances, the pattern was clearly another image and the young draftsman was to rely on the skill of others to perfect his own technique. A would-be artist must first become a good judge of prints before becoming a successful draftsman.

"Nature" was the ultimate pattern of all patterns for drawings of the human body, and in the eleventh, and final, use of the term, nature was the pattern to be followed:

> ...yet must neither the one nor the other be drawne in such a posture as will not agree with the motion of nature, that is, to draw such a posture which a man cannot imitate with his naturall body, and so for any thing else whatsoever nature must be the patterne of all kind of drafts (14).

The student was warned that the various poses of the body must be kept in mind when drawing the human figure. Nature must be followed; fancy must not be given free reign. As Abraham Cowley put it in his poem, "To the Royal Society," a draftsman "must not...content himself to make it like

/ Th'Ideas and the Images which ly / In his own Fancy, or his Memory."[36] This emphasis on nature being the ultimate guide to creating a work of art was tied to Dürer's own insistence that nature must be the first source for inspiration and not other artists' works.[37]

The student was encouraged to study the work of a master, before nature, because the master had done this for him. "Famous Masters'" works were to be studied as they themselves had studied all types of characters "to the end they might express them to the life in their drawings and works" (12). Diligent observation of masters' works was a necessary precursor to the direct observation of nature. Being a "famous Master" then was dependent on developing keen observational skills and learning to draw was dependent on following others' drawings.

A Book of Drawing provided both visual and written guidance in the art of drawing. Plates were used in conjunction with text to help the student be fruitful in his pursuit. The method depended on repeated copying of "the works of famous Masters" before proceeding to draw from nature itself. The author realized that the young get impatient and are reluctant to keep practicing. His message was introduced with an opening Latin tag on the title page about the joys of learning and this moral was matched with a closing rhyming precept on the necessity for patience and diligence to achieve skill in draftsmanship:

> *Be patient, thou that seekest for this skill,*
> *By grace and art so mayst thou have thy will* (40).

A Book of Drawing aimed to provide the young gentleman "the Ground-work to make him fit for doing any thing by hand." The text strove to help them recognize "good Masters workes" and improve their own ability to judge these from the work of "botchers." Readers were being taught to both consume and produce images, while developing their own skills in both fields.

"Draw and Engrave their Schemes with Delight and Assurance"

The emphasis in the previous section on the application of techniques and practices of judgment involved in selecting models and producing copies leads to the next element involved in the visual education of natural

36 Cowley, "To the Royal Society," in Sprat, *History of the Royal Society*, B2r.
37 Parshall, "Graphic Knowledge," 396.

philosophers, namely a sustained engagement with and knowledge of the history of engraving as well as with the authority vested in its principle artists. In the dedicatory letter at the beginning of *Sculptura* addressed to Robert Boyle (1627–1691), another founding member of the Royal Society, John Evelyn made clear that he was publishing this book at Boyle's urging and that he did so to encourage others to study the topic:

> ...as you [Boyle] are pleased to judge it useful for the encouragement of the Gentlemen of our Nation who sometimes please themselves with these innocent diversions (Collections worthy of them for divers respects) and, especially, that such as are addicted to the more Noble *Mathematical Sciences*, may draw, and engrave their *Schemes* with delight and assurance, I have been induc'd to think it more worthy your Patronage, and of my small Adventure... (A2v)

In this dedication, Evelyn set up the text as a practical manual by holding out the possibility that natural philosophers might engrave their own figures and aligned it with the Royal Society's Baconian goal of creating a history of trades. To that end, Fellows of the Royal Society read the book. For instance, Francis Willughby recorded passages from it in his commonplace book.[38] The bulk of the text that follows, however, is neither a technical manual on the intricacies of the chemical properties of *aqua fortis* nor a discussion of the finer points of manually cutting lines into a prepared copper plate; instead, Evelyn went to great lengths to explore the history of engraving both linguistically and from the point of view of connoisseurship. The premise of the book then is not to provide a textbook for making a print, but rather a textbook for recognizing a good print. The assumption at the core of the book is that you need to train your eye before you can proceed to training your hand.

Best known to historians for his detailed diary, John Evelyn was also a prolific author and translator.[39] As Michael Hunter has shown in his explorations of Evelyn's early years through his letters and other manuscript sources, Evelyn's staunch Royalist leanings and the English Civil War pushed him toward connoisseurship.[40] Although he saw public service as the proper pursuit for a man of his upbringing, after the execution of Charles I in 1649 there was no possibility of him following that path, so while living

38 Poole, "The Willughby Library," 238.
39 Chambers, "Evelyn, John."
40 Hunter, "John Evelyn," 67–98.

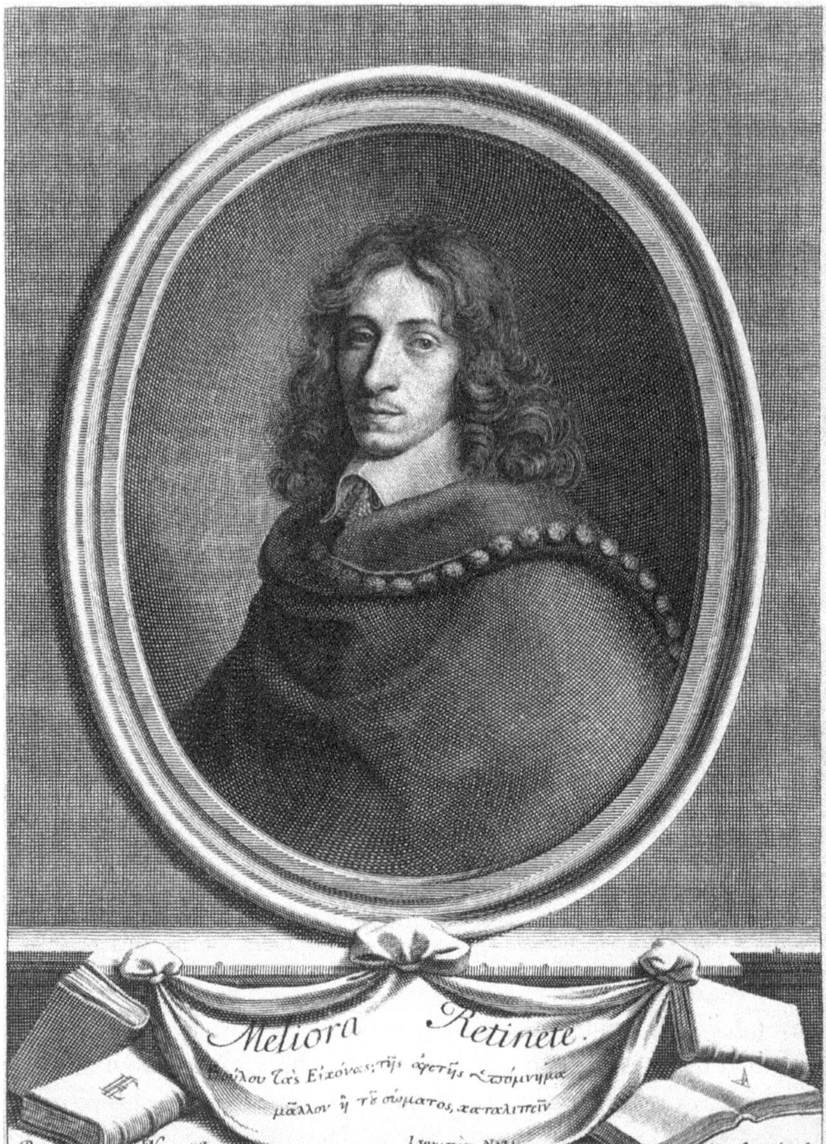

Figure 1.8: Robert Nanteuil, *Portrait of John Evelyn*, Engraving, 1650. National Gallery of Art, Washington, D.C., Rosenwald Collection, 1943.3.6509.

in Paris beginning in 1649 he devoted himself to the study of the arts. He had shown an interest in the arts early; however, his pursuit of them had been discouraged by his grandfather as well as his tutor.[41] In addition to

41 Hunter, "John Evelyn," 71.

collecting objects, he also collected information from artists and developed friendships with them.

Abraham Bosse and Robert Nanteuil were among the artists Evelyn visited and from whom he learned about drawing and printmaking.[42] Evelyn viewed Bosse as both a source of prints and of instruction. It was during this time that Evelyn sat for a portrait by Nanteuil, which captures his interest in prints and books more broadly. (Fig. 1.8) A portion of Evelyn's motto *Omnia explorate; meliora retinete* (explore everything; keep the better) fills the space where the sitter's name usually goes. Although taken from a biblical verse, his motto, particularly the portion on the print, *meliora retinete*, encapsulated the central tenet of Evelyn's growing interest in connoisseurship.[43]

While on his Grand Tour in Italy in 1645, Evelyn made some of the first known landscape drawings by an English amateur.[44] Although his drawings do not display a great deal of skill, they do serve as a testament to Evelyn's belief in the ability of images to convey the sense of a place. To that end, in 1649, while living in Paris, he created a suite of etchings based on his own drawings made while he was in Italy. (Fig. 1.9) Like the drawings, the prints do not exhibit a high degree of technical skill. What I want to focus on is not Evelyn's skill, or lack thereof, but rather his decision to make these prints at all. While making a drawing can be understood in the context of Baldassare Castiglione's directions for the successful courtier, taking the next step of making a series of etchings is outside of what most other gentlemen were doing then.[45] Although Evelyn mentioned his personal connections to artists in his discussions of their work in *Sculptura* and he noted Hevelius' image-making skill, he did not mention his own attempts. While Evelyn's omission of his own etchings could be read as an ambivalence towards producing prints, it should be read instead as an exercise in his own visual judgment, as the prints were not mentioned because they should not be considered a master's work, sufficiently expert for others to follow.

Although he had originally intended to publish a translation of Abraham Bosse's (1604–1676) treatise on engraving and etching, which was the first such text, *Sculptura* includes very little technical information.[46] In a letter to John Wilkins, Evelyn wrote about setting aside his translation of Bosse's

42 Hunter, "John Evelyn," 68. Kusukawa also discusses his connections to Nanteuil and Bosse. Kusukawa, "Early Royal Society," 363.

43 The motto is taken from I Thessalonians 5:21.

44 Sloan, *'A Noble Art,'* 20.

45 On Castiglione's legacy in England see Bermingham, *Learning to Draw*, esp. chap. 1, "Drawing and the Courtly Art of Illusion," 3–32.

46 Bosse, *Traicté*. Evelyn, *Sculptura*, An Advertisement.

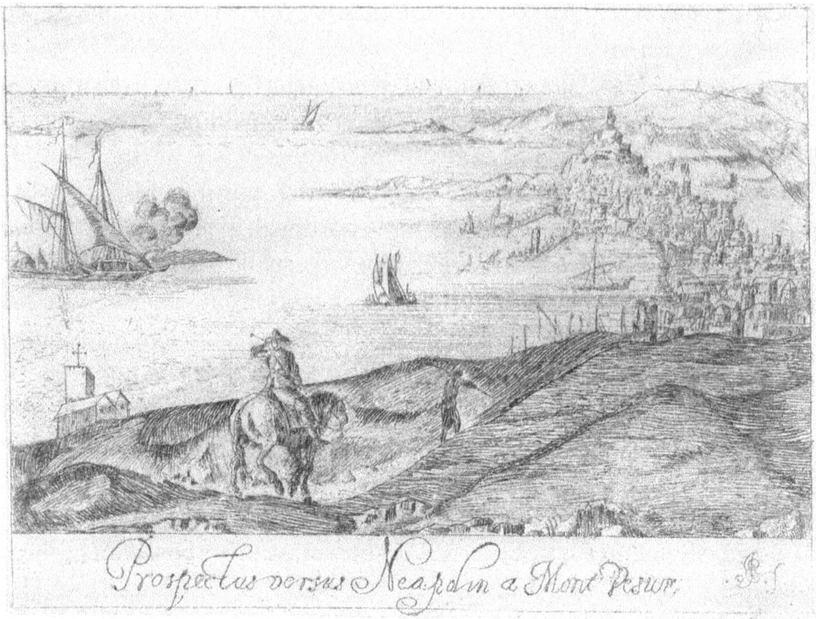

Figure 1.9: John Evelyn, *View of Naples from Mount Vesuvius*, Etching, 1649, 105 x 141 mm (trimmed). The British Museum, 1838,0530.1.

text because he had happened into William Faithorne's shop and learned that Faithorne, "himself a skillful graver," had completed such a translation and it was in fact already printed and about to be published.[47] Even if Evelyn thought this section was the "most useful" of his intended book, he did not present his text as a window into a craftsman's workshop as had previous books on the arts such as John Bate's *The Mysteryes of Nature and Art* (London, 1634).[48] The illustrated title page of Bate's book featured eight small vignettes which gave a taste of what was promised in the four sections of the text: waterworks, fireworks, drawing, painting and engraving, and "divers Experiments." (Fig. 1.10) William Eamon noted the popularity of Bate's book and its place among a growing number of publications aimed at revealing "secrets" to the virtuosi.[49]

In contrast to Bate's title page, Evelyn's frontispiece, which he designed, presented his readers with a muse-like representation of engraving. (Fig. 1.11)

47 John Evelyn to John Wilkins, January 29, 1660/61, Archives of the Royal Society, Early Letters, EL/E/1; and Letter Book Original, LBO/1/49.

48 Evelyn to Wilkins, January 29, 1660/61.

49 Eamon, *Science and the Secrets of Nature*, 307.

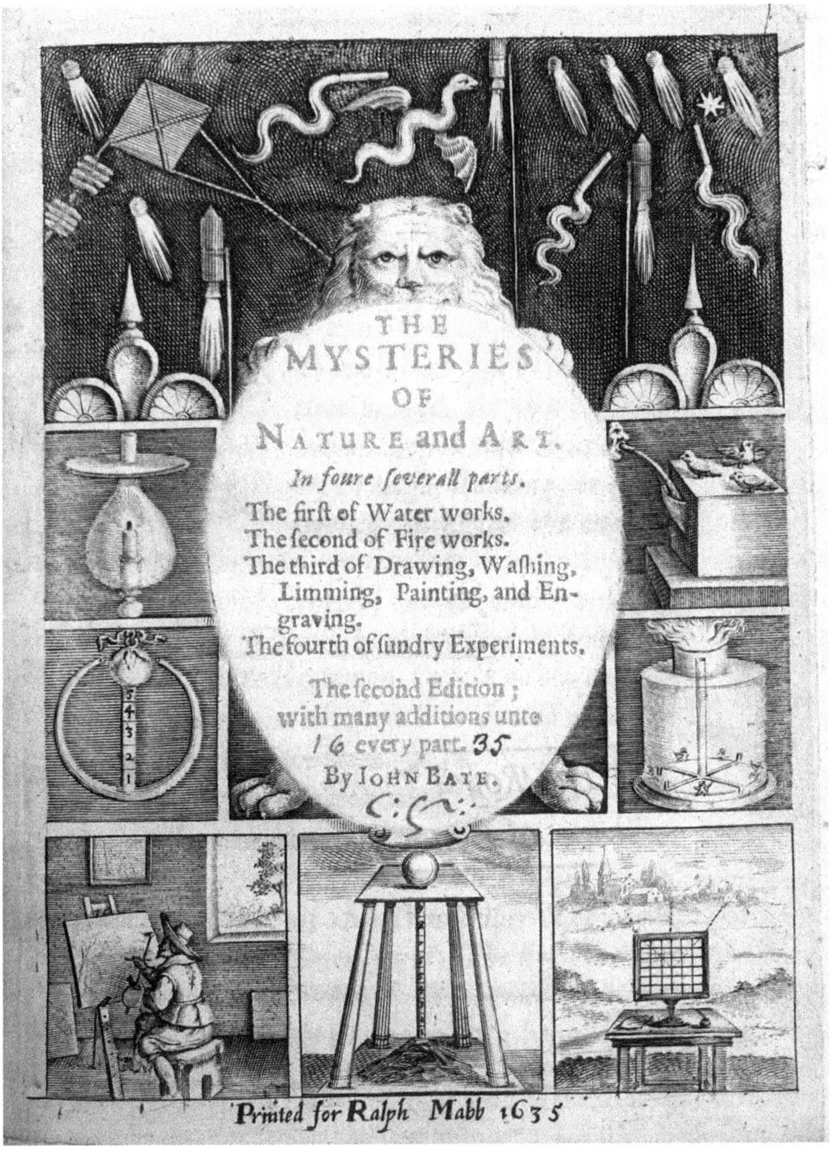

Figure 1.10: Anonymous, *Title page*, Engraving, from John Bate, *The Mysteries of Nature and Art* (London: Printed for Ralph Mab, 1635). Courtesy of The Royal Society.

The meta-textual image is striking: a classically draped woman surrounded by an engraver's tools sits outdoors preparing a plate for the engraved title page for Evelyn's book. Although Evelyn did not mention his early landscape drawings and etchings, he did clearly pronounce his authorship of this image by signing it in the lower left, "J E inve": and by including his initials on the title page the muse holds. The essence of Evelyn's intellectual project is here encapsulated in an image made after his own manual labor. The intellectual and the manual are thus inextricably linked, modeling the practices he proposed for "the Gentlemen of our Nation." The text, which followed this classicized frontispiece, proceeded in a similarly humanist manner beginning with a lengthy exploration of the etymology of sculpture and its derivatives.[50] The first three chapters of the book work to establish a long and noble history for the arts that led to the development of engraving and etching, his main focus, in order to substantiate his later claim for their inclusion among the liberal arts.[51]

The first hundred pages of the book are dedicated to a detailed exploration of the history of various forms of cutting lines into surfaces, providing both a context for current printmaking practices and establishing printmaking as an appropriate pursuit for the "Gentlemen of our Nation." Creating a scholarly lineage for the practice of engraving begins with a detailed etymology of sculpture and its derivative practices and terms. Not surprisingly, this exploration involves passages in Latin, Greek, and Hebrew. This application to visually focused materials of the central humanist practices of philology and source study also represents an innovation in the shift from traditional humanism toward the more pragmatic approaches natural philosophers were taking in their studies of craft and trade. At the beginning of his book, Abraham Bosse traced the history of his subject briefly back to before Moses.[52] Evelyn went much further both in terms of chronology and depth. Casting aside the notion that "this Art had its being from Eternity," he traced the origins of sculpture to the infancy of letters and therefore to Adam (11). He went on to discuss antediluvian accounts of the carving of texts into hard surfaces. Evelyn admitted that some of his evidence was quite arcane and he did not to want to "put too much stresse upon superannuated Tradition," and so he turned to Moses:

> …this we are sure is of *Faith*, and without controversy; That in *Moses* we have the *Tables of stone*, engraven by the *Finger of GOD* himself; where the commandement is expresse, even against the abuse of this very *Art*, as well as an instance of the Antiquity of Idolatry attesting that of Sculpture: THY SHALT NOT MAKE TO THY SELF ANY GRAVEN IMAGE (15).

50 This combination of literary style with practical information fits with Michael Hunter's description of Evelyn's writing style in *Sylva* and Hunter's characterization of *Elysium Britannicum* as "unashamedly élitist." Hunter, "John Evelyn," 95.

51 Evelyn, *Scupltura*, 132.

52 "…ni qu'il est des plus anciens puis que Moyse en a escrit ainsi que d'une chose laquelle estoit de son temps fort en usage." Bosse, *Traicté*, 1.

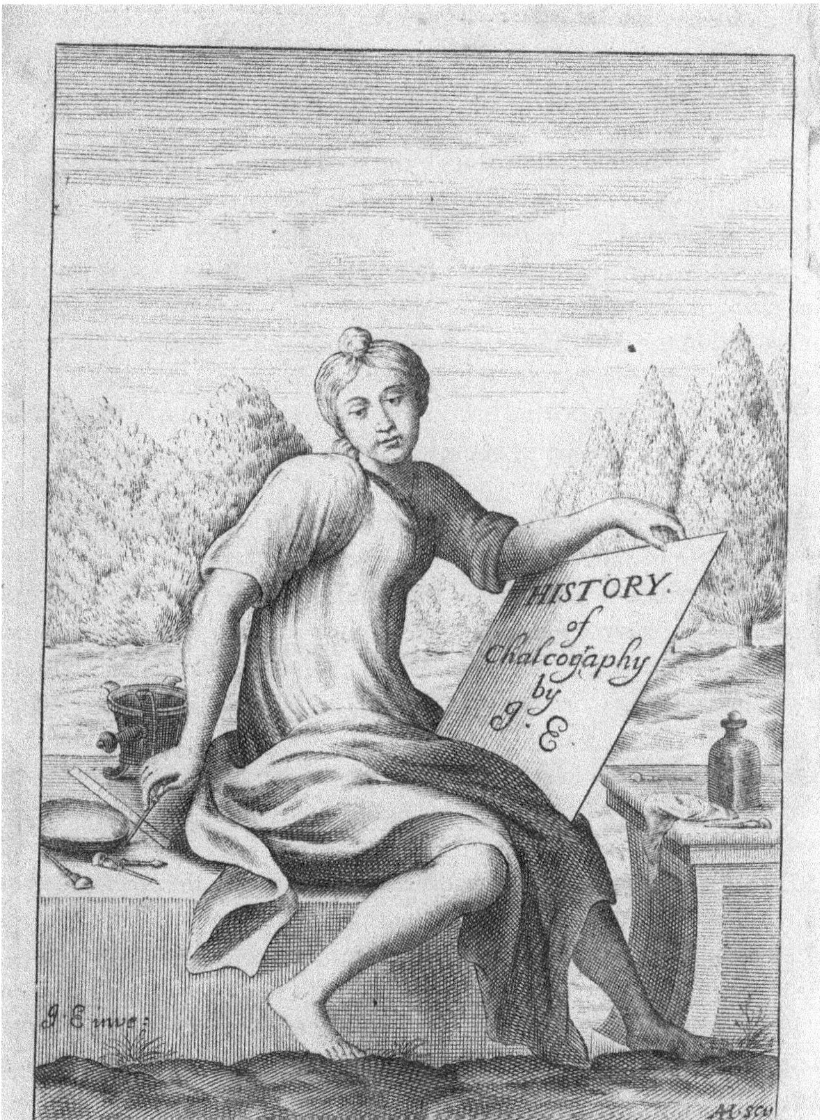

Figure 1.11: John Evelyn, *Frontispiece*, Etching, from *Sculptura* (London: Printed by J. C. for G. Beedle & T. Collins, 1662). The Bodleian Libraries, University of Oxford, 8vo G.119 Linc.

The irony of this commandment does not seem lost on Evelyn as he highlighted the use of "this very *Art*" in its prohibition. Furthermore, recalling God's creation of the Commandments through the process of engraving positions it as an epistemic method, one which records authoritative knowledge, rather than solely an aesthetic one. Evelyn provided a detailed, humanistic history lesson to establish the legitimacy of engraving as a practice and pastime for gentlemen and to elevate it out of the confines of the artist's studio and placing it within the realm of the natural philosopher.

Throughout the text, Evelyn reinforced the corollary necessity of developing your judgment through the study of good models. Once the long and illustrious prehistory of etching and engraving was established, Evelyn proceeded to recount the great masters of these arts, as well as occasionally mentioning those who made woodcuts. As Craig Hanson has pointed out *Sculptura* "provided for its original readers not only an introduction to prints but a framework for understanding the history of European art more generally."[53] Every artist mentioned is followed by a list of exemplary works. During his time in exile, Evelyn amassed a sizeable print collection.[54] The listing of the best prints by each master was included so the reader could seek out good models with which to train his eyes. At points, Evelyn gave brief synopses of why a particular artist's work was to be valued. For instance, in discussing Francesco Villamena (ca. 1565–1624), Evelyn described him as "a rare workman, whether consider'd for the equality of his hatches, which he conducted with a liberty and agreeableness suitable to the perfection of his design" (55) drawing particular attention to his engraving of Christ in the Temple. This passage combines Evelyn's brief account of the valuable characteristics of Villamena's work with a concrete example, which provided his readers with guidance for their own looking as well as a piece of evidence upon which they could test their own looking process.

Lest Evelyn's readers think that finding good models to copy was something left to beginners, he provided multiple examples throughout the text of well-known artists using prints as models for their own compositions. He drew attention to this historical practice to reinforce the importance of it in learning to draw and engrave: "Nor let the most supercilious *painter* despise what we have here alledged; or imagine it any diminution to his *Art*, that he now and then put his hand to the *pen*, and draw even after some of the *Masters* we has so much celebrated" (117). For example, Evelyn mentioned the anecdote (related by Giorgio Vasari in the *Lives of the Painters*) of Michelangelo copying an engraving by Martin Schongauer of "Saint Anthonies Temptation" at a young age (37).[55] He later described how

53 Hanson, *The English Virtuoso,* 83.

54 Hunter, *Image of Restoration Science*, 40.

55 The painting in the collection of the Kimbell Art Museum: Michelangelo Buonarroti, "The Torment of Saint Anthony," Tempera on panel, 1487, AP 2009.01.

Andrea del Sarto based some of his work on prints by Albrecht Dürer (117). These reminders of how famous artists also copied prints helped further to establish the authority of Evelyn's method. By referencing good models, Evelyn provided a method whereby his readers could develop their own visual judgment. The collection of models supported the emulation of these masters. He set down a type of Baconian program for the acquisition of a particular kind of knowledge, that is you must examine and collect particular instances, compare and differentiate them, and you will arrive at truth.

"A fit subject for our kingdomes knowledge and practice"

A shift in the gentlemanly practice in visual arts from one centered on delight to an approach rooted in a practical pedagogy reflects the emergence of a new mode of visual judgment heralded by the application of this technique for looking at and generating images. This shift also tracks the maturation of an increasingly regular pedagogy that the self-taught natural philosopher could pursue confidently. In *Sculptura* John Evelyn promoted the study of the history of engraving as well as the practice of it and insisted that his readers "not only to be well skill'd in their way of design; but to be able also to perform something in the Art themselves" (62). Despite this exhortation, he provided very little in the way of practical instruction in *Sculptura*. Evelyn had wanted to publish a translation of Abraham Bosse's *Traicté des manières de graveur en taille douce*, but as we saw, he set aside his project after learning that William Faithorne had already completed a similar project. Where Bosse was writing his manual to shore up the place of intaglio printing in the hierarchy of the arts, Faithorne was not concerned with the academic standing of engraving in England.

Like Evelyn, Faithorne spent time in Paris in exile and brought an awareness of the state of French art back to London with him. Horace Walpole, in his *A Catalogue of Engravers, Who have been Born, or Resided in England*, viewed Faithorne's time in France as crucial to his success:

> There had been one Englishman, who without the timid perfection of the French masters, had shown that softness and force, freedom and finishing, were compatible, and that the effect of chiaro scuro did not depend on unblended masses of white and black. This was William Faithorne.[56]

56 Walpole, *Catalogue*, as quoted in: Griffiths, *Print in Stuart England*, 126.

For Walpole then, Faithorne's exposure to French art allowed him to better grasp how to render objects in space and how to give a sense of light hitting objects. It was during this time that Faithorne met Abraham Bosse. They had shared concerns about the regulations placed on printmakers and printsellers in Paris, as is evidenced by their mutual involvement in a protest against new taxes and licensing regulations.[57]

Faithorne's translation of Bosse's text, published in 1662, appears to have been the first treatise on engraving to be published in English.[58] The differences in motivations come through most clearly in their paratexts. As discussed earlier, in his preface Bosse set up engraving and etching as ancient arts that played an important role in human history since Moses came down from the Mount.[59] Faithorne, on the other hand, quickly dismissed the need to justify his choice of occupation.[60] Rather than providing a justification of his art, Faithorne was concerned "to present my Countrey with something of use, profit, and delight" (Av). Faithorne's book was grounded in an English context, and he gave instructions that were specific to English artists.[61] Instead of using his text to aggrandize the place of engraving in the hierarchy of the Arts, as Bosse did, Faithorne used his work for the benefit of his countrymen and to educate them about his art.

In his dedicatory letter, Faithorne made clear that his book was aimed at those who wished to both understand and make engravings. He stated that he decided to publish the book "because it [graving and etching] hath arrived to such an height in these our latter times, as it becomes a fit subject for our kingdomes knowledge and practice" (A2r). As with the text in *A Book of Drawing* and *Sculptura*, Faithorne worked to develop the visual judgment of his readers. He did this through describing the best practices of his art and indicating particular areas where practitioners were apt to struggle. In doing so, Faithorne provided descriptions of visual markers of bad workmanship with directions for how to correct the work. Although Faithorne's text contained far more technical information than Evelyn's

57 Marianne Grivel found a document that includes both Bosse and Faithorne's signatures on a petition protesting François Mansart's 1651 proposed tax on prints. Grivel, *Commerce de l'Estampe*, 94–96.

58 An expanded discussion of this was previously published in Doherty, "Creating Standards," 15–36.

59 Bosse, *Traicté*, "Avant Prospos," 1.

60 Faithorne, *Art of Graveing*, A2r.

61 Regarding the "Englishness" of Faithorne's book, for instance, at the beginning of the section on "How to planish and polish your Plate" he states: "Here in *England* you must buy your Cooper ready forged from the Brasiers." Faithorne, *Art of Graveing*, 4.

and focused primarily on manual skills, he still made a concerted effort to train the eye of his reader alongside his hand. Readers were taught how to recognize their own errors as well as how to correct them. The consumption and production of visual materials were central to his goals in publishing.

In addition to its stated goal of providing information about how to make an intaglio printed image, Faithorne's book also improved the likelihood that an image would create the illusion of accuracy. As we saw with his very different images of women discussed in the Introduction, accuracy was context specific. At the same time by openly discussing standardized processes for making engravings, Faithorne allowed his readers to better understand the illusion. The word "accurate" occurs once in Faithorne's text and is closely aligned with the importance of "neatness," and both are understood as the results of great care, tying his usage to those of Fellows of the Royal Society whose texts introduced the word into the English language:[62]

> Be carefull that your fingers do not interpose between the plate and the Graver, for they will be troublesome, and hinder you in carrying your Graver level with the plate, so that you cannot make your strokes with that freedome and neatnesse, as otherwise you may. This I think fit to give you notice of in this place, because the skill of holding your Graver is that which you must first perfectly learn, and be able to practise without pain or difficulty; or else you will not gain so great a readinesse and command of your hand, as is required in an accurate and skilfull Graver (45–46).

I have quoted this passage at length to give a sense of the range of meanings associated with accuracy in this text and the importance of methodical work in the achievement of success. In this passage, the term "graver" is used to refer to both the tool used by an engraver and as a metonym for the engraver himself. This ambiguity reinforces the sense that the reliability and truth of testimony were related to an individual and, in this case, one who was trained to be accurate and skillful. This collapse of the artisan and his tool allows us to understand *The Art of Graveing and Etching* as a manual for making and being.

As the first text in English to outline a detailed method for engraving and etching, Faithorne's *The Art of Graveing and Etching* provided readers with a set of best practices to be followed to achieve a desired outcome.

62 See the Introduction for a further discussion of the etymology of "accuracy." *Oxford English Dictionary*, s.v. "accurate," "accuracy," "accurately," and "accurateness."

These best practices established the possibility that an engraver could reproduce a design as it was given to him and served as the large-scale structure for explicating a method for producing an engraving or etching that had the visual traces of accuracy. As laid out by Faithorne, the key components for judging the quality of one's own work included: 1) the use of conditional statements to clarify how to know if a best practice had been properly implemented; 2) the inclusion of cautionary asides which outlined how the engraver might go astray in the process of making an engraving; and 3) the framing of descriptions of best practices with references to the plates included in the book, which also serve as models. These three components worked to regulate the actions of the engraver, train his eye, and standardize his output. Furthermore, they provided a clear guide for using the burin which was similar to Robert Hooke's outline of the use of the microscope. Both were instruments which were of use to Fellows of the Royal Society and both required study and patience to learn to use well.

Reading Faithorne's text alongside those of Robert Boyle affords a better understanding of what it meant to advocate for the development of readers' ability to judge their own work in the second half of the seventeenth century. Boyle is an exemplary figure because of his noted commitment to experimentation, which is borne out in his published works, as well as for the prominent role he played as a gentleman natural philosopher in mid-century London.[63] Boyle professed a preference for a plainer style of writing, to more clearly inform his readers.[64] In addition to being a founding member of the Royal Society, Boyle wrote many texts promoting experimentation and explained in great detail how to conduct experiments.[65] Boyle's writings,

63 The importance of Robert Boyle to the history of the Royal Society and the development of modern science has been made clear in the work of Steven Shapin, Simon Schaffer, and especially Michael Hunter. See for example: Shapin, *Social History of Truth*, where Boyle is set up as the quintessential gentleman-philosopher; Shapin and Schaffer, *Leviathan and the Air-Pump*; and the writings of Michael Hunter, in particular, *Robert Boyle (1627–1691)* and *Robert Boyle Reconsidered*.

64 "And first, as for the style of our Experimental Essaies, I suppose you will readily find that I have endeavour'd to write rather in a Philosophical than a Rhetorical strain, as desiring that my Expressions should be rather clear and significant than curiously adorn'd.... And certainly these Discourses, where our Designe is only to inform Readers, not to delight or perswade them, Perspicuity ought to be esteem'd at least one of the best Qualifications of a style." Boyle, *Certain Philosophical Essays* (London, 1661), 11. John Harwood writes about the balance Boyle tried to find in his writing between plain style and the tools of classical rhetoric he thought were useful for engaging readers. Harwood, "Science Writing and Writing Science," 46.

65 See for example: Boyle, *New Experiments,*; Boyle, *Certain Physicological Essays*; Boyle, *Some Considerations Touching*; and Boyle, *Continuation of New Experiments*.

like Faithorne's, stressed the necessity of careful and precise actions to ensure accuracy in the methods they espoused. Studying Boyle alongside Faithorne allows us to see how Faithorne established precision in rendering as a condition of accuracy.

In addition, Robert Boyle had connections to William Faithorne. In autumn 1664, William Faithorne was working on a portrait of Robert Boyle.[66] (Fig. 1.12) It was among the many Faithorne did of Fellows of the Royal Society and was just one of the projects that tied him to the Royal Society.[67] In addition to a traditional three-quarter bust, the portrait includes a rendering of one of Boyle's air-pumps. The portrait of Boyle, Boyle's writings, and Faithorne's writings all serve as testaments to the importance of accuracy within the network of individuals working in and around the Royal Society, as all three visually and verbally embody the importance of careful work in the creation of knowledge.

In Robert Boyle's writings in *Medicina Hydrostatica, or Hydrostaticks Applyed to the Materia Medica* (1690) on the experiments he conducted to determine the specific gravity of minerals in order to correctly identify them, conditional statements played a key role in guiding the experimenter in setting up his balance and achieving precise measurements. Boyle gave an overview of the experimental setup; outlining the best practices for setting up the balance in particular. After reviewing a couple of details of the setup, Boyle used the conditional to warn his readers about potentials for error and the importance of following his prescribed method: "…if any of these Circumstances be not taken care of, (as it happens, when we are not heedful enough) the true Weight of the Solid is somewhat altered."[68] Not being "heedful enough," both as a reader and as an experimenter, would lead to imprecise results, which in this case means that you would not have correctly measured the specific gravity of your sample and therefore you would misidentify it, possibly leading to a grave error in prescribing the medicinal use of the substance. In this example, close attention to the production of the experimental setup was essential to studying the natural world, and Boyle's writing style worked to control the actions of the experimenter in order to attempt to provide reliable testimony. Rose-Mary Sargent has written about Boyle's diffidence towards theorization and his

66 Based on the correspondence between Robert Boyle and Robert Hooke, acting as Boyle's agent, we can date this portrait to the fall of 1664. *Correspondence of Robert Boyle*, vol. 2, 304, 316, 412, 442.

67 For a list of Faithorne's portrait drawings, see: Bell and Poole, "English Seventeenth-Century Portrait Drawings," 53–54.

68 Boyle, *Medicina Hydrostatica*, 16.

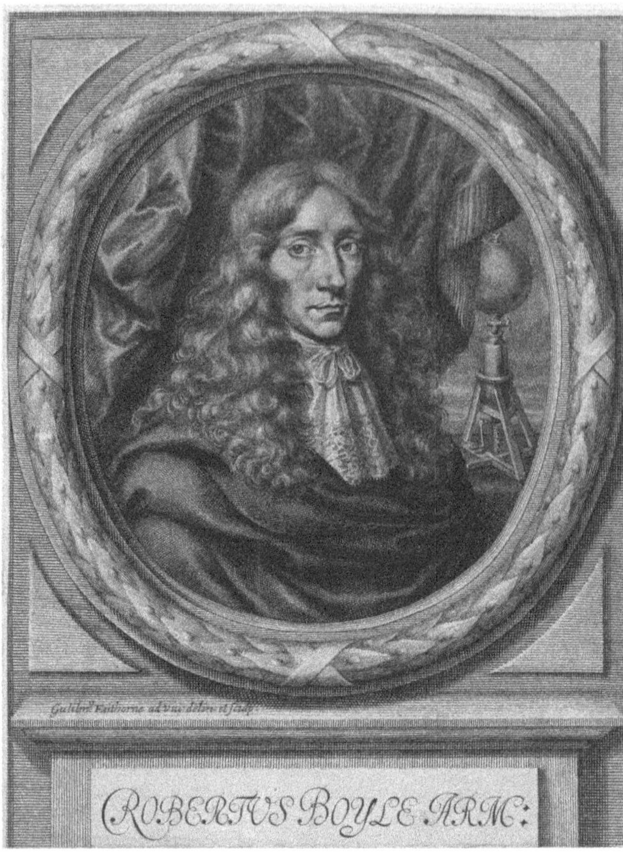

Figure 1.12: William Faithorne, *Portrait of Robert Boyle*, Engraving, 1664. National Portrait Gallery, London, NPG D22648.

goal in his writing of having his data and method be clear. As she put it, "he was not seeking a reputation as an ingenious theorist but was trying to lay the foundation for a new way of doing science."[69] This "new way" shared stylistic conventions with Faithorne's way of writing.

One of the rhetorical and practical situations in which conditional statements have a great deal of impact on the accuracy of an engraver's work was in the whetting of the burin or needle. These statements identified a problem and offer that the solution lay in properly sharpening the instrument. Although these statements were not always set off with if/then clauses, the sense was generally: "If you find that your point cuts not freely and smoothly, 'tis because it is not whet exactly round (12)." Recognizing a flawed

69 Sargent, *Diffident Naturalist*, 185.

line and correcting your tools were both visual and manual processes. Here the implication was that if you want your needle to move smoothly through the varnish on the plate, then you must whet it "exactly round." This level of precision was particularly important in the case of an etching created to accompany natural history texts, as stray lines or ones that do not curve in the way intended can detract from the truth-value of the images. Faithorne had previously described the best practices for whetting the etching needles "exactly round" and this if/then statement provided guidelines for gauging how well the rules for whetting were followed.

This pair of examples give a sense of how Boyle and Faithorne used conditional statements to clarify and augment the best practices they outlined by developing care in the reader to follow them. These statements provide a structure for judging whether or not the best practices had been put into use well or whether some adjustments were necessary. While it was possible to make an etching without heeding the nuanced advice given in these conditional phrases, the resulting print was less likely to turn out *exactly* as you intended. Not following the method provided and heeding the advice extended the distance between the depiction of an object and the object itself. Faithorne's method worked to decrease this distance and to allow for the possibility of producing accuracy in your work.

Another technique that Boyle and Faithorne used to ensure the faithful execution of the best practices was to provide their readers with cautionary asides that highlighted moments in the process where they might go astray. The cautionary aside played a critical role in the language Robert Boyle used in his writings on constructing an air-pump as it was essential to the success of his experiments that the vacuum chamber have a tight seal. Boyle used the aside to indicate to the reader a place in the process where errors might occur. For example, in describing how to construct the receiver of the pump, Boyle cautioned the reader to be aware of potential difficulty: (Fig. 1.13)

> The last thing belonging to our Receiver, is the stop-cock designed in the first Figure by (N.) for the better fastening of which to the neck, and exacter exclusion of the Air, there was soder'd on to the shank of the Cock (X) a Plate of Tin, (MTUW) long enough to cover the neck of the Receiver. But because the cementing of this was a matter of some difficulty, it will not be amiss to mention here the manner of it...[70]

70 Boyle, *New Experiments*, 11. The parenthetical letters in this passage refer to letters in the accompanying plate.

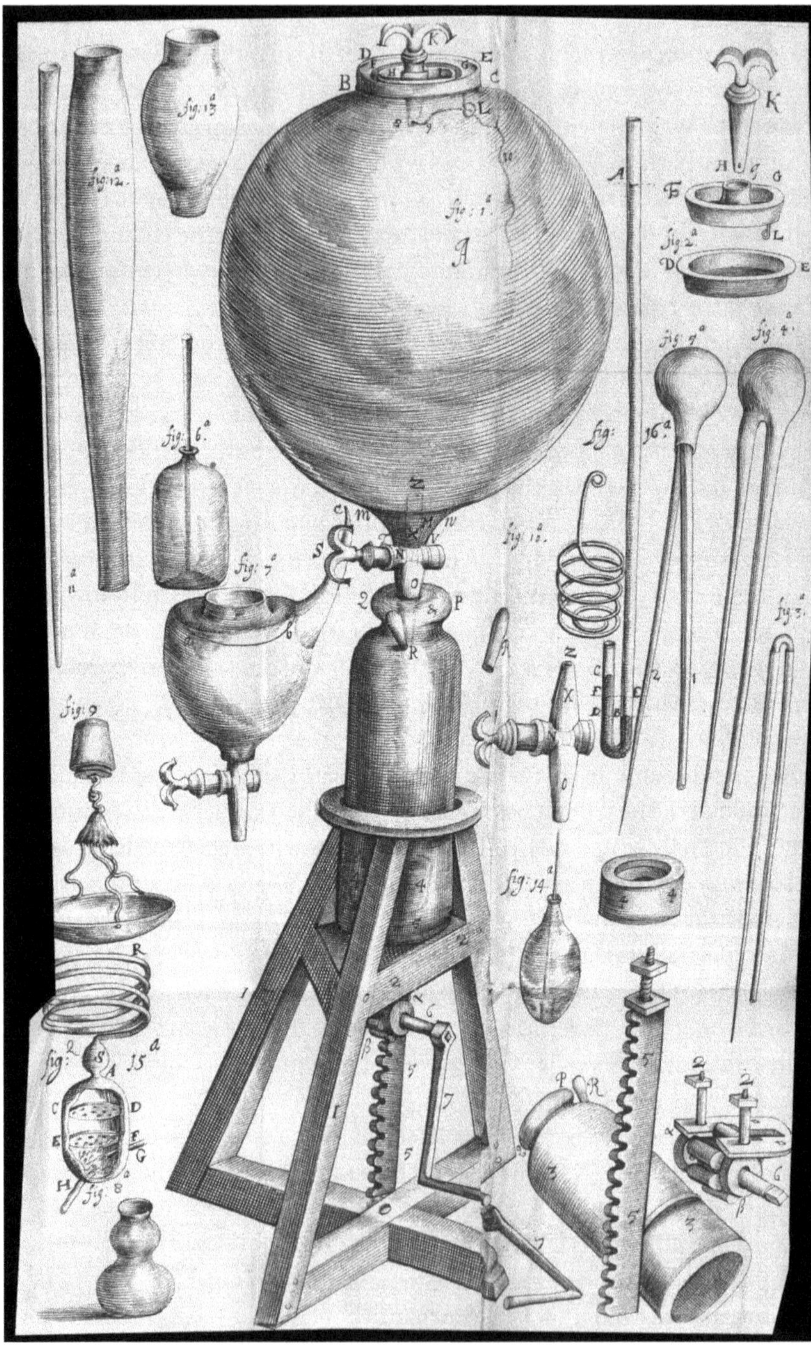

Figure 1.13: Anonymous, *Air Pump*, Engraving, from Robert Boyle, *New Experiments Physico-Mechanicall, Touching the Spring of the Air, and Its Effects* (Oxford: Printed by H: Hall for Tho: Robinson, 1660). © The Royal Society.

Boyle then continued with a highly specific description of how to avoid the difficulty that he had in constructing his pump. Boyle here described, in general, how to assemble this section of the air-pump and then proceeded to refine his method by drawing on his own practical experience.

Drawn from Faithorne's own experience as well as Bosse's, the asides in *The Art of Graveing* raised practical concerns about points at which error could enter the process and helped the reader avoid mistakes in his attempts to put Faithorne's prescribed method into practice. Within the section titled "How to apply your hard Varnish on the Plate, and make it black," Faithorne delineated the method for applying the resist to the plate and included a cautionary aside to ensure the proper application.

> Take great care, that there be not too much Varnish upon the plate, and that your hand be not sweaty; because the sweat mixing with the Varnish, wil cause little bubbles, when it is applied to the fire, which will become little holes in the Varnish (7).

Two concerns were raised in the passage, the first dealt with the amount of varnish applied and the second dealt with the hands of the artisan. Both cautions were outside of the realm of explicating method, but both helped to clarify and further protect against stray marks on the plate. This aside provided a more nuanced understanding of the best practices for applying varnish by drawing on Faithorne's own practice as an engraver. These asides worked to further codify the standards that Faithorne provided his reader.

The third element that developed reader's judgment was the framing of best practices with references to the images interleaved with the text. Boyle's *New Experiments Physico-Mechanicall* framed descriptions of best practices with reference to images to help clarify the steps to be taken. In this text on experimental natural philosophy, the images included modeled the ideal construction of the experimental set-up. Although there is only one plate accompanying the text, it is made up of a series of figures, each with its own set of labels. (See Fig. 1.13) The first reference to this figure occurred on page nine as Boyle tried to clarify the shape of the glass that would serve as the vacuum chamber in the pump. The text gave little sense of the ideal shape: "[The glass vessel] consists of a Glass with a wide hole at the top, of a cover to that hole, and of a stop-cock fastened to the end of the neck, as the bottom."[71] In order to know that the vessel should be roughly oval in shape and taper at the end that will connect to the stop-cock, it is

71 Boyle, *New Experiments*, 9.

necessary to refer to the provided figure. This detailed image is covered with letters and numbers that key into the accompanying text and each line was precisely cut into the copper to ensure clear communication so that the reader would better understand how to construct an air-pump.

The 10 plates in *The Art of Graveing and Etching* were all engraved by William Faithorne. He referred to these figures throughout the text.[72] References to the figures in the text performed a similar function as the other two techniques discussed, as they worked to clarify the reader's understanding of the method and further ensured the reader's ability to replicate the method in his own work. Again, the images were based on Faithorne's own experience as he had to put his own method into action to create these prints that translated his verbal method into pictorial evidence. If this sounds like circular logic that is part of the point; the plates and the text were both the products of Faithorne's years of practical experience as an engraver and worked to reinforce his credibility as someone capable of providing a method worth following. The relationship between image and text re-inscribed the collapse of the engraver and his tools and further stressed the need for careful actions.

For instance, the text facing the last plate began with a reference to the image and continually referred the reader back to the image. (Fig. 1.14) Section 27, "The manner how to hold your Graver, with other particulars," began by directing the reader to the facing image: "You may see also that the uppermost part of this figure describes to you the form of the two Gravers, with their handles fitted for the whetting" (45). Two sets of information were to be gleaned from the upper portion of this image at this moment. First, the reader was shown another view of what the square and "lozeng" points of a burin look like. These two possibilities for the point were shown in the previous image from a different angle. Second, this section of the image shows how engravers often cut off half of the handle to make sure that the knob did not rub against the plate. The figure both provided more detailed information than the text and modeled behavior because in this case we can assume that Faithorne used a burin that had been cut in the manner he advocated and because the plate itself was free of imperfections.

Although the reader was directed to the image to learn about these two aspects of preparing the burin for engraving, the reason for cutting the handle was only given after the reference to the image:

72 "Figure(s)" occurs thirty-three times in the book indicating that the individual images were referred to multiple times.

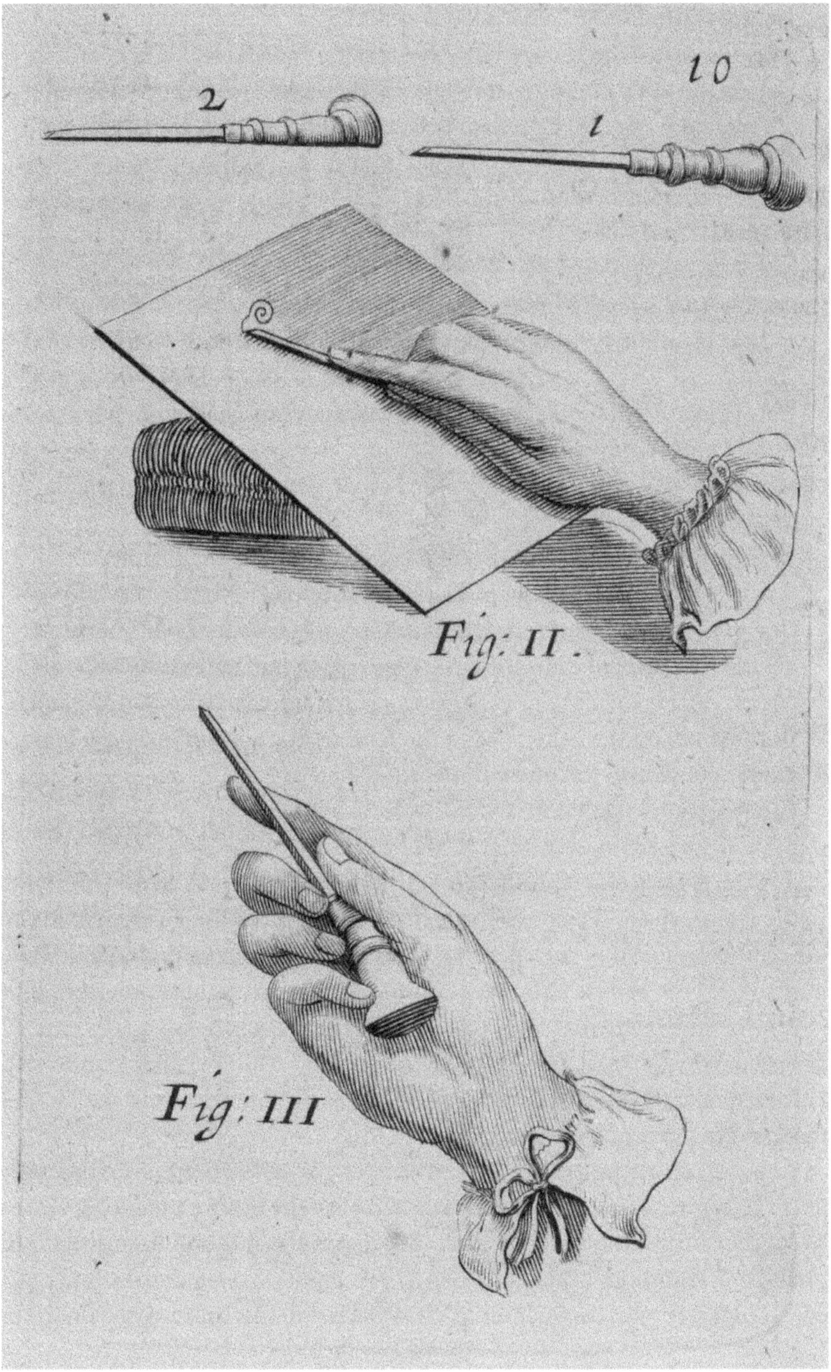

Figure 1.14: William Faithorne, *Plate 10*, Engraving, from *The Art of Graveing and Etching* (London: Published by the Author, 1662). The Newberry Library, Case Wing Z 417 .282.

> They that use this Art, do before they make use of them, commonly cut away the part of the knob or bowl that is at the end of their handles, which is upon the same line with the edge of their Graver; to the end it may not obstruct or hinder them in their graving, as the figure II. shews you (45).

This additional information clarified what was seen in the upper portion of the image and was coupled with a reference to another part of the same image. The first paragraph of this section began and ended by directing the reader to the facing image. The best practices for preparing your burin relied on visual information to convey clearly the how and why of the method.

Now that the "other particulars" associated with holding the burin were textually and visually explained, the third paragraph of this section and the third portion of the image dealt more directly with holding the burin:

> The third figure describes to you the way of holding the Graver; which is in this manner. You must place the knob or ball of the handle of your Graver in the hollow of your hand, and having extended your forefinger towards the point of your Graver, laying it opposite to the edge that should cut the copper, place your other fingers on the side of your handle, and your thumb on the other side of the Graver, in such sort that you may guide your Graver flat and parallel with the plate; as you may see in the IIII. [III] figure (45).[73]

Again, the description of the best practices began and ended with references to the facing image. The reader was to first look at the image, then read the description, then look back at the image to be sure that he understood how best to hold the burin. This visual reinforcement of the best practices that Faithorne outlined worked to ensure that his readers developed a careful practice. The plates in *The Art of Graveing and Etching* visually embodied Faithorne's exposition of the ideal method of producing an engraving and modeled behavior for his readers.

In both Boyle's writings and Faithorne's *The Art of Graveing*, the references to the accompanying plates worked with the other elements outlined to further refine the reader's attempts to put the method described into practice. The authors' asides and uses of the conditional regulated the actions of the engraver or experimenter in order to ensure that an accurate image or

73 As there are only three sections to this image, we can assume that "IIII." is a typographical error and III is intended.

an air-pump with a tight seal was created. By publishing his book, Faithorne both sought to help his reader be "an accurate and skilfull Graver" and provided insight for "all Lovers of Art" to understand how intaglio-printed images were created.[74]

Conclusion

The three books discussed in detail in this chapter did not exist in isolation. They were read, studied, and used.[75] Fellows of the Royal Society were part of this group of readers and users. The effects of accuracy that this group of users helped to develop into a epistemically replete approach to images include the developments I have tracked across this chapter, particularly the development of visual judgment; an engagement with proper models, tools, and techniques of image reproduction; the application of a visual vocabulary formed by line and paper to translate three-dimensional space into two-dimensional renderings; an understanding of the history of images and its construction of authority; and the internalization of practical models of instruction as well as the contexts for learning how to create images. Allow me to conclude by offering one further and brief example that synthesizes these elements. In addition to owning or reading all the books discussed in this chapter, Robert Hooke also put pen to paper in the manner prescribed.[76] (Fig. 1.15) As will be discussed further in chapter two, Hooke was deeply involved with the production and consumption of images. While there are very few figural studies in his extant papers, this collection of sketches resembles the disparate groups of heads and profiles brought together in the plates of works like *A Book of Drawing*. (Fig. 1.16) Hooke's sketches combined profile outlines with more detailed costume studies. The drawings show how the continual addition of lines could create the illusion of a three-dimensional body in two dimensions. Both parallel lines and hatching were used to give this sense of weight and depth. In this drawing, the methods provided by *A Book of Drawing* were put into practice by a Fellow of the Royal Society.

74 Faithorne, *The Art of Graveing*, 46 and A2v.

75 For an example of a drawing manual with clear signs of use see for instance: *Diagraphia, sive ars delineatoria* tracing of recto on verso of plate eight. National Art Library, UK, Special Collections, II.RC.D Box I.

76 Hunter, "Hooke's Figurations," 251–260.

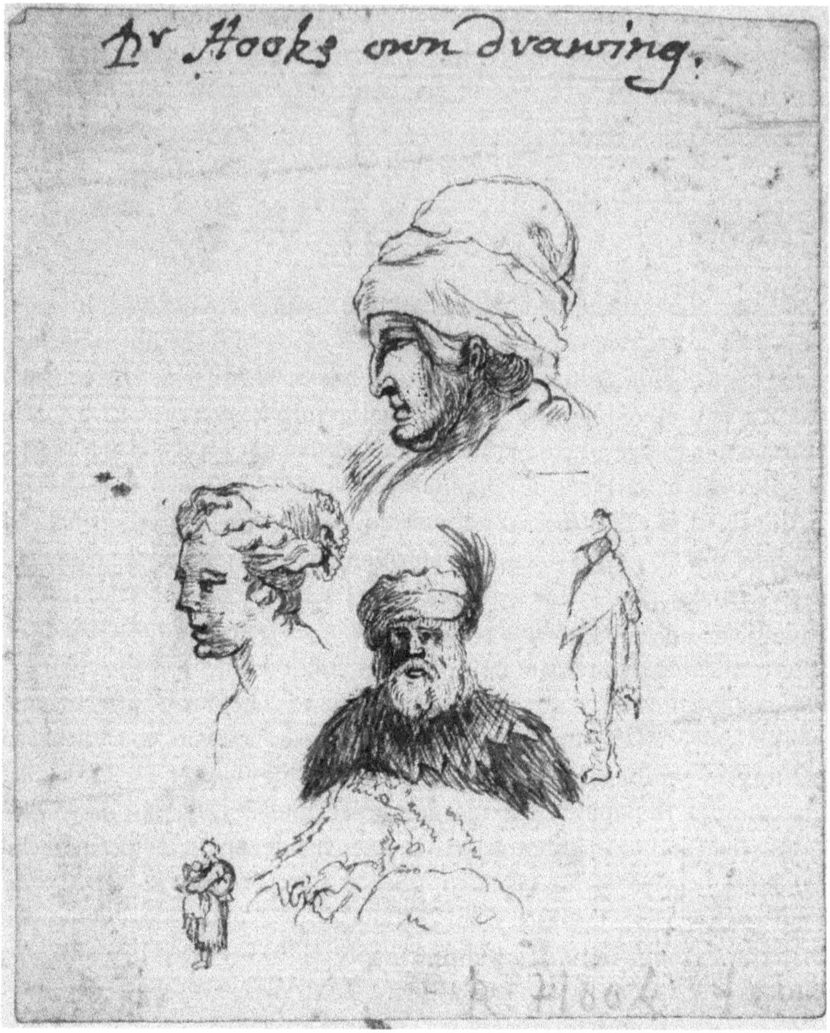

Figure 1.15: Robert Hooke, *Three Heads and Two Figure Studies*, Ink on Paper, c. 1660. Tate Britain, T10678.

Knowing how the image-making process worked let viewers better understand the "innocent *Witch-craft* of lights and shades," which in turn helped them consent to the illusion of accuracy.[77] The three books examined in this chapter together pulled back the curtain on this magic. *A Book of Drawing*, with Albrecht Dürer looking on from the beginning of the book, provided access to a method that had name recognition while

77 Evelyn, *Sculptura*, 106.

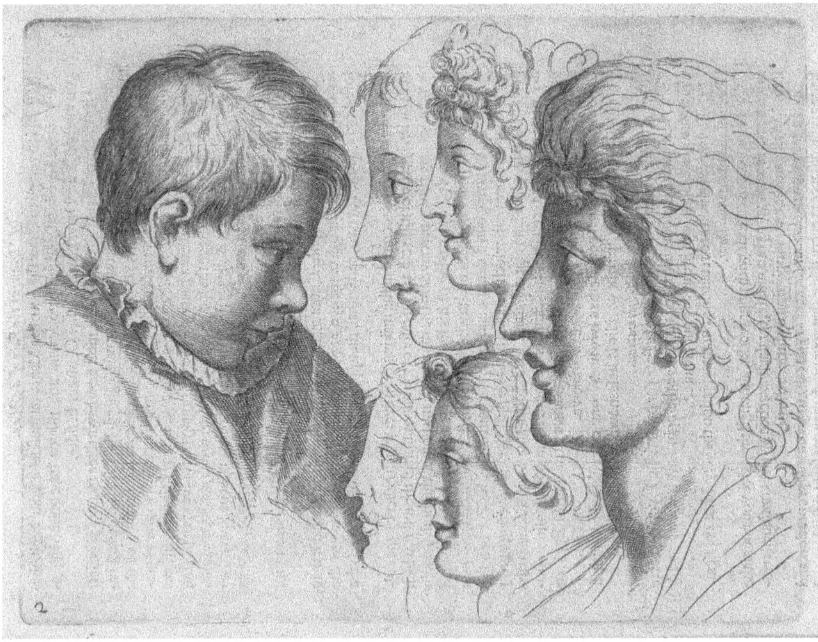

Figure 1.16: Anonymous, *How to draw faces*, Engraving, from *A book of drawing, limning, washing or colouring of maps and prints, or, The young-man's time well spent* (London, Royal Exchange: Printed by James and Joseph Moxon, for Thomas Jenner, 1647). Yale Center for British Art, Paul Mellon Collection, NC730 .B6 1647+ Oversize.

at the same time exposing an English audience to the main tenets of Continental art practice. The emphasis on "good Masters workes" trained the eye of the reader while they sought to train their hands. John Evelyn's text reinforced the importance of good models and developed his reader's skills in connoisseurship while framing his text as being useful to "such as are addicted to the more *Noble Mathematical Sciences.*" At the same time, Faithorne presented readers with a text Evelyn intended to publish. Evelyn yielded his own project, acknowledging the value of Faithorne's practical knowledge of the subject and clear skill with it. Without the context of *A Book of Drawing* and *Sculptura* though, *The Art of Graveing* is just a technical manual; with that context in place, it helped shift the entire grounds of the study and visual representation of the natural world toward accuracy. Examining and understanding these texts in the broader context of the early Royal Society allows us to see that accuracy was itself a dynamic concept.

Bibliography

I. Manuscript Sources

London, Archives of the Royal Society

EL/E. Early Letters. Letters of John Evelyn.
EL/H2. Early Letters. Letters of Johannes Hevelius.
LBO/1. Letter Book Original. 1661–1666.
LBO/2. Letter Book Original. 1667–1668.

London, British Library

Add. MS 78340. Evelyn Papers. Vol. CLXXIII. Collection in two parts, the first of medical and culinary recipes, the second of skills and crafts; 1659–[early 18th cent?].

II. Printed Primary Sources

Academia Italica, The Publick School of Drawing of the Gentlemans Accomplishment. London: Printed by P. Lillicrap, 1666.

Bate, John. *The Mysteryes of Nature and Art: Conteined in Foure Severall Tretises, the First of Water Workes, the Second of Fyre workes, the Third of Drawing, Colouring, Painting, and Engraving, the Fourth of Divers Experiments, as wel Serviceable as Delightful.* London: Imprinted for Ralph Mab, 1634.

A Book of Drawing, Limning, Washing or Colouring of Maps and Prints, or, the Young-Man's Time Well Spent. London: Printed by James and Joseph Moxon, for Thomas Jenner, 1647.

A Booke of Drawinges. Performed according to the best order for use & Brevity that is yet Extant. London: Printed and are to be sould by Peter Stent, 1650.

A Booke of Portraicture. London: Sold by Godfrey Richards, 1665.

Bosse, Abraham. *Moyen Universel de Practiquer la Perspective sur les Tableaux, ou Surfaces Irregulieres.* Paris: Chez ledit Bosse, 1653.

—. *Traicte des Manieres de Graver en Taille Douce sur l'Airin.* Paris: Chez Bosse, 1645.

Boyle, Robert. *Certain Physiological Essays, Written at Distant Times, and on Several Occasions.* London: Printed for Henry Herringman, 1661.

—. *A Continuation of New Experiments, Physico-Mechanical, Touching the Spring and Weight of the Air, and Their Effects.* London: Printed by H. Hall for R. Davis, 1669.

—. *The Correspondence of Robert Boyle.* Edited by Michael Cyril William Hunter, Antonio Clericuzio and Lawrence Principe. 6 vols. London: Pickering & Chatto, 2001.

—. *Medicina Hydrostatica, or, Hydrostaticks Applyed to the Materia Medica.* London: Printed for Samuel Smith, 1690.

—. *New Experiments Physico-Mechanicall, Touching the Spring of the Air, and Its Effects, Made, for the Most Part, in a New Pneumatical Engine.* Oxford: Printed by H: Hall for Tho: Robinson, 1660.

—. *Some Considerations Touching the Usefulnesse of Experimental Naturall Philosophy.* Oxford: Printed by Henry Hall, 1664.

Browne, Alexander. *Ars Pictoria: Or an Academy Treating of Drawing, Painting, Limning, and Etching.* London: Printed by J. Remayne for the Author, 1669.

—. *A Compendious Drawing-Book.* London: Printed for Austin Oldisworth, 1677.

—. *The Whole Art of Drawing, Painting, Limning, and Etching.* London: Printed for Peter Stint, 1660.

Cousin, Jean. *Livre de Pourtraittre de Maistre Iean Cousin Peintre et Geometrien Tres-Excellent.* Paris: Chez Jean le Clerc, 1595.

Diagraphia, sive Ars Delineatoria. Amsterdam: Joannes Janssonius, 1616.

Dürer, Albrecht. *Underweysung der messung mit dem zirckel un richt scheyt.* Nuremberg: I. Formschneyder, 1525.

Evelyn, John. *The Diary of John Evelyn.* Edited by E.S. de Beer. Oxford: Clarendon Press, 1955.

—. *Sculptura: Or the History, and Art of Chalcography and Engraving in Copper.* London: Printed by J. C. for G. Beedle & T. Collins, 1662.

The Excellency of the Pen and Pencil, Exemplifying the Uses of them in the most Exquisite and Mysterious Art of Drawings, Etching, Engraving, Limning. Painting in Oyl, Washing of Maps and Pictures. London: Printed by Thomas Ratcliff and Thomas Daniel, for Dorman Newman and Richard Jones, 1668.

Faithorne, William. *The Art of Graveing, and Etching, Wherein Is Exprest the True Way of Graveing in Copper.* London: Published by the Author, 1662.

Hevelius, Johannes. *Selenographia; Sive, Lunae Descriptio.* Gdansk: Autoris sumtibus, Typis Hünefeldianis, 1647.

Hooke, Robert. *Micrographia, or, Some Physiological Descriptions of Minute Bodies Made.* London: Printed by J. Martyn and J. Allestry, 1665.

Oldenburg, Henry. *Correspondence of Henry Oldenburg.* Edited by A. Rupert Hall and Marie Boas Hall. Madison: University of Wisconsin Press, 1965.

Peacham, Henry. *The Art of Drawing with the Pen, and Limning in Watercolours.* London: Richard Braddock for William Jones, 1607.

Pepys, Samuel. *The Diary of Samuel Pepys: A New and Complete Transcription.* Edited by Robert Latham and William Matthews. Berkeley: University of California Press, 1970.

Sprat, Thomas. *The History of the Royal-Society of London for the Improving of Natural Knowledge.* London: Printed by T. R. for J. Martyn and J. Allstrey, 1667.

III. Secondary Sources

Bartrum, Giulia. *Albrecht Dürer and His Legacy: The Graphic Work of a Renaissance Artist*. Princeton, N.J.: Princeton University Press, 2002.

Barzman, Karen-edis. *The Florentine Academy and the early modern state: The discipline of disegno*. Cambridge: Cambridge University Press, 2000.

Bermingham, Ann. *Learning to Draw: Studies in the Cultural History of a Polite and Useful Art*. New Haven: Yale University Press, 2000.

Bohlin, Diane DeGrazia. *Prints and Related Drawings by the Carracci Family: A Catalogue Raisonné*. Washington, D.C.: National Gallery of Art, 1979.

Bubenik, Andrea. *Reframing Albrecht Dürer: The Appropriation of Art, 1528–1700*. Surrey, UK: Ashgate, 2013.

Chambers, Douglas D. C. "Evelyn, John (1620–1706)." *Oxford Dictionary of National Biography* (Oxford: Oxford University Press, Sept 2004).

Doherty, Meghan C. "Creating Standards of Accuracy: Faithorne's *The Art of Graveing* and the Royal Society," in *Science in Print: Essays on the History of Science and the Culture of Print*. Edited by Rima D. Apple, Gregory J. Downey, and Stephen L. Vaughn. Madison: University of Wisconsin Press, 2012, 15–36.

—. "*The Young-Mans Time Well Spent*: Learning to Draw from a Master," in *Publishing the Fine and Applied Arts, 1500–2000*, edited by Michael Harris, Giles Mandelbrot, and Robin Meyers. London: British Library Press, 2012, 51–78.

Eamon, William. *Science and the Secrets of Nature: Books of Secrets in Medieval and Early Modern Culture*. Princeton: Princeton University Press, 1994.

Flick, Gert-Rudolf. *Masters & Pupils: The Artistic Succession from Perugino to Manet, 1480–1880*. London: Hogarth Arts; Paul Holberton Publishing, 2008.

Fowler, Caroline O., *Drawing and the Senses: An Early Modern History*. Turnhout, Belgium:

Harvey Miller Series in Baroque Art/Brepols, 2017.

Fransen, Sietske and Katherine M. Reinhart. "The Practice of Copying in Making Knowledge in Early Modern Europe: An Introduction." *Word & Image* 35, no. 3 (2019): 211–222.

Goldstein, Carl. *Teaching Art: Academies and Schools from Vasari to Albers*. Cambridge: Cambridge University Press, 1996.

Griffiths, Antony. *The Print in Stuart Britain, 1603–1689*. London: Published for the Trustees of the British Museum, 1998.

Grivel, Marianne. *Le Commerce de l'Estampe à Paris au XVIIe Siècle*. Geneva: Librarie Droz, 1986.

Hanson, Craig Ashley. *The English Virtuoso: Art, Medicine, and Antiquarianism in the Age of Empiricism*. Chicago: University of Chicago Press, 2009.

Harwood, John T. "Science Writing and Writing Science: Boyle and Rhetorical Theory." In *Robert Boyle Reconsidered*, edited by Michael Cyril William Hunter. Cambridge: Cambridge University Press, 1994.

Henderson, Felicity. "Robert Hooke and the Visual World of the Early Royal Society." *Perspectives on Science* 27, no. 3 (May–June 2019): 395–434.

Hults, Linda C. *The Print in the Western World: An Introductory History*. Madison: University of Wisconsin Press, 1996.

Hunter, Matthew C. "Hooke's Figurations: A Figural Drawing Attributed to Robert Hooke." *Notes and Records of the Royal Society* 64, no. 3 (2010): 251–260.

Hunter, Michael Cyril William. *The Image of Restoration Science: The Frontispiece of Thomas Sprat's* History of the Royal Society *(1667)*. London: Routledge, 2017.

—. "John Evelyn in the 1650s: A Virtuoso in Quest of a Role." In *Science and the Shape of Orthodoxy: Intellectual Chance in Seventeenth-Century Britain*, 67–98. Woodbridge: The Boydell Press, 1995.

—. *Robert Boyle, 1627–91: Scrupulosity and Science*. Woodbridge, Suffolk: Boydell Press, 2000.

—. *Robert Boyle Reconsidered*. Cambridge: Cambridge University Press, 1994.

Lincoln, Evelyn. *The Invention of the Italian Renaissance Printmaker*. New Haven: Yale University Press, 2000.

Kusukawa, Sachiko. "The Early Royal Society and Visual Culture." *Perspectives on Science* 27, no. 3 (May–June 2019): 350–394.

MacGregor, William B. "The Authority of Prints: An Early Modern Perspective." *Art History* 22, no. 3 (1999): 389–420.

Parshall, Peter. "Graphic Knowledge: Albrecht Dürer and the Imagination." *Art Bulletin* 95, no. 3 (2013): 393–410.

Poole, William. "The Willughby Library in the Time of Francis the Naturalist." In *Virtuoso by Nature: The scientific worlds of Francis Willughby FRS (1635–1672)*, edited by Tim Birkhead, 227–243. Leiden: Brill, 2016.

Sargent, Rose-Mary. *The Diffident Naturalist: Robert Boyle and the Philosophy of Experiment*. Chicago: University of Chicago Press, 1995.

Shapin, Steven. *A Social History of Truth: Civility and Science in Seventeenth-Century England*. Chicago: University of Chicago Press, 1994.

Shapin, Steven, and Simon Schaffer. *Leviathan and the Air-Pump: Hobbes, Boyle, and the Experimental Life*. Princeton, N.J.: Princeton University Press, 1985.

Shuffelton, George. *Codex Ashmole 61: A Compilation of Popular Middle English Verse*. The Camelot Project at the University of Rochester. Kalamazoo, Mich.: Medieval Institute Publications, 2008.

Sloan, Kim. *'A Noble Art': Amateur Artists and Drawing Masters, c.1600–1800*. London: British Museum Press, 2000.

Walpole, Horace. *A Catalogue of Engravers, Who Have Been Born, or Resided in England*. 2d ed. London: Printed for J. Dodsley, 1786.

2. "A New Visible World": Developing a Visual Vocabulary for the Microscopic

Abstract
This chapter uses Robert Hooke's *Micrographia* to examine the intersection of visual conventions in portraiture with the viewing of the microscopic world. In the "Preface" to *Micrographia*, Hooke asserted that he had discovered "a new visible World" through the help of newly invented optical devices. Before the publication of *Micrographia* there was little visual consensus about how best to display this new microscopic world. For Hooke, accuracy was produced through repeated looking and drawing with continued reference to the visible world. Hooke was aided by the visual vocabulary developed by engravers for translating a three-dimensional world into a two-dimensional representation of it, and his awareness of these conventions is what set his illustrations apart from his predecessors.

Keywords: Robert Hooke, Microscope, Portraiture, Engraving

Robert Hooke (1635–1703) saw differently.[1] Not only that, he described what he saw in a way that others could understand with precise reference to what he saw and the visual culture with which he and his readers were familiar.[2] When he looked through a microscope, he saw what others did not or could not and he translated that perception with the aid of his broad visual education.[3] For example, when John Wilkins, a founding Fellow of

1 An earlier version of this chapter appeared as: Doherty, "Discovering the 'True Form'," 211–234.

2 Felicity Henderson has also discussed how Hooke saw differently with specific reference to his methodological writings. Henderson, "Robert Hooke," 398–400.

3 While historians of science may not find Edgerton's thesis about the geometrization of space compelling, the connections he makes between Galileo and Florentine *disegno* are informative for my own approach to Hooke and seventeenth-century visual culture. Edgerton, *Heritage of Giotto's Geometry*, 223–253. For an example of a contemporary finding Galileo's view of the moon transformative of his own viewing see: Bloom, "Borrowed Perceptions," 117–122.

Doherty, M.C., *Engraving Accuracy in Early Modern England: Visual Communication at the Early Royal Society*. Taylor & Francis Group, 2022.
DOI 10.5117/9789463721066_CH02

the Royal Society of London, looked through a microscope at deer hair, he saw a quill-like structure. Hooke, on the other hand, saw a sponge-like form. Hooke was right.[4] This chapter explores how seeing differently, in this case, was facilitated by Hooke's early training and lifelong interest in the arts, which developed his visual judgement. Hooke's understanding of line and particularly the engraved line allowed him to resolve the images he saw through his microscope in ways that others could not.[5]

Hooke described his working method in the Preface to *Micrographia:* "And therefore I never began to make any draughts before by many examinations in several lights, and in several positions to those lights, I had discover'd the true form."[6] His working method was critical to his seeing differently. Hooke struggled with discovering the "true form" of his object of study and this awareness, coupled with his familiarity with artistic techniques, allowed him to see the microscopic world differently than his predecessors and contemporaries. Before the publication of *Micrographia* there was little visual consensus about how best to display this new microscopic world. For Hooke, accuracy was produced through repeated looking and drawing with continued reference to the visible world.

At a meeting of the Royal Society in 1663, George Ent related to the Fellows that deer skins laid in the bottom of a boat would tend to float above the other animal skins laid with them when submerged in water.[7] This remark led John Wilkins to relate his own observations of deer's hair with a microscope: he "found them to be tubulous throughout, like quills; whereas the hair of other animals was like canes, full of pores, but not hollow; so that deer's hair being filled with more air than that of other animals, endeavoured, when put under water, to emerge to the top."[8] Hooke was then "desired to observe in his microscope some of the deer's hair, what cavities they have."[9] Although Wilkins was a founding member of the Royal Society and had served as an early patron to Hooke, his opinion was not sufficient proof for the Fellows that deer's hairs were in fact "tubulous." Instead they called upon Hooke's expertise. Two weeks after this first mention of the nature of deer's hair, "Mr. Hooke produced a microscopical observation of the hair

4 Examinations of deer hair with modern microscopes reveal the sponge-like form of the hair.

5 For a discussion of the optical issues that complicated the use of microscopes in the seventeenth century, see: Turner, "Impact of Hooke's *Micrographia*," 124–125.

6 Hooke, *Micrographia*, "The Preface," f2v.

7 Birch, *History of the Royal Society*, vol. 1, 342 (9 December 1663).

8 Birch, *History of the Royal Society*, vol. 1, 342 (9 December 1663).

9 Birch, *History of the Royal Society*, vol. 1, 343 (9 December 1663).

of an Indian deer, which represents it to be like a spunge, not like quills."[10] The sponge-like form of the hair referred to its interior structure; it does not have a perforated exterior like a sponge. Hooke closely examined the sample of deer hair provided and then produced a drawing that allowed the Fellows to see what he saw.

While the drawing Hooke showed at the meeting is not extant, he did include a description and image of deer hair in *Micrographia*. (Fig. 2.1) In the accompanying text, he further explicated his findings and the steps he took to reach his conclusion about the sponge-like nature of the hair: "through the Microscope, it appears all perforated from side to side, and Spongie, like a small kind of spongy Coral, which is often found upon the English shores; but though I cut it transversly, I could not perceive that it had any pores that ran the long-way of the hair."[11] To help his readers better understand what he saw when looking through his microscope, he compared it directly with another natural object they could see without the aid of the microscope. Although Wilkins and Hooke would have been using comparable instruments, Hooke saw something different and this ability set him apart from the other Fellows.

Among the many legacies Hooke left to the Royal Society's pursuits, his development of a set of shared conventions for and approaches to the transmission of the information he discerned through visual study at the microscope stands as an under-appreciated contribution to early modern science. Hooke's interest in producing accurate images is paired with his desire to ensure their proper reception and interpretation. This twin motivation led him to develop several shared visual conventions: making explicit the steps entailed in the physical preparation of specimens for the microscope; the communication of light and shadows, material slices, and sample depths in his printed images; the repeated practices of looking and the reproduction of similar visual effects across different specimens and even different categories of visual objects.

By looking repeatedly at the same object, and by cutting and recutting his specimen, Hooke was able to see what others could not and thereby discovered the "true form." Hooke's examination of deer hair validated the underlying assumption of the flotation of deer hair while clarifying why it was actually buoyant. The Council of the Royal Society repeatedly directed Hooke, in his role as Curator of Experiments, to examine objects that were brought to meetings and to subsequently bring drawings made after his microscopic

10 Birch, *History of the Royal Society*, vol. 1, 348 (23 December 1663).

11 Hooke, *Micrographia*, 158.

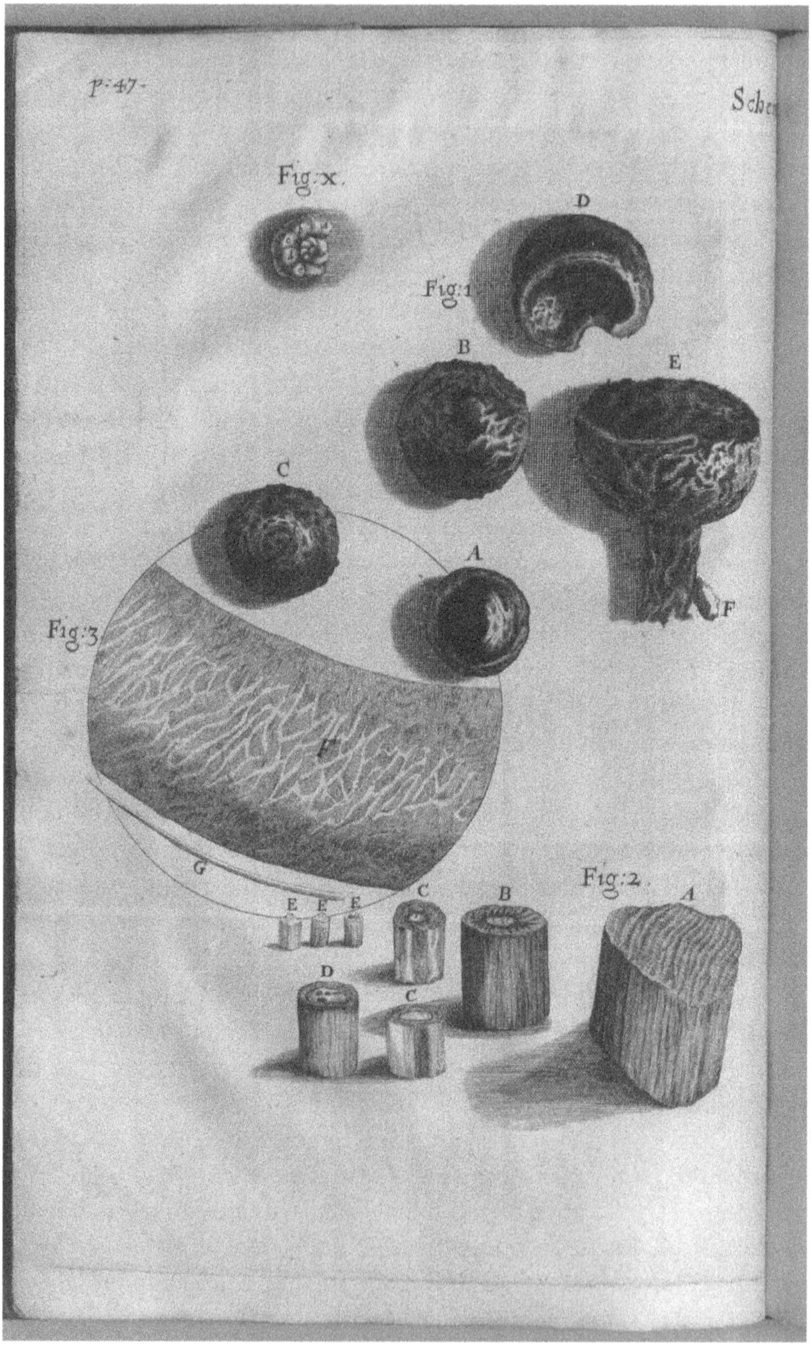

Figure 2.1: Anonymous, *Scheme 5*, Engraving, from Robert Hooke, *Micrographia* (London: Printed by J. Martyn and J. Allestry, 1665). By courtesy of the Department of Special Collections, Memorial Library, University of Wisconsin-Madison, Gift of Daniel and Eleanor Albert, Oversize RE26 O62 H66 M53 1665.

observations to meetings of the Royal Society.[12] As will be discussed in more detail later, these images, as well as his printed ones, were notably different from what was produced by earlier users of the microscope. By arguing that Hooke saw differently, I am not however suggesting that he was exceptional, instead I submit that he had developed a new way of seeing, that is, a new visual methodology, and that he wanted to instruct others to follow his method.

In the Preface to his *Micrographia*, Robert Hooke asserted that he had discovered "a new visible World" through the help of newly invented optical devices that added "*artificial Organs* to the *natural*."[13] Hooke was further aided in his establishment of a new visual method by the visual vocabulary developed by engravers for translating a three-dimensional world into a two-dimensional representation of it. In particular this chapter brings together evidence from two seemingly disparate bodies of work: seventeenth-century portrait engravings and the thirty-eight plates that illustrate Hooke's *Micrographia*. This juxtaposition of images from different genres highlights the debt Hooke owed to the visual vocabulary of engravers in his attempts to discover the "true form" of his object of study and how his ability to see differently was tied to his engagement with mid-seventeenth-century visual culture.

My use of the term vocabulary here connotes the visual elements used by engravers to create the illusion of three dimensions in a two-dimensional medium, the same "innocent *Witch-craft* of lights and shades" that concerned John Evelyn, as discussed in chapter one. This illusion of three dimensions is produced through the manipulation of depth and contrast, which were paramount concerns for Hooke. That is, the lines carved by the engraver created a sense of dimensionality both in the forms depicted and in the space they inhabit by manipulating the viewer's perceptions. Lines were used to alter how the viewer perceived the illusion of depth within the picture plane and how light struck different areas within the space of the image in different ways to provide contrast. This chapter explores how these terms of the visual vocabulary of engraving, while not fully developed in the early visual culture of microscopy, are mobilized in the illustrations in Hooke's *Micrographia* to create a compelling view of "a new visible World." Accuracy in this context meant that Hooke's images looked more like portraits than like previously published microscopic images and were therefore more readily accepted by his viewers as presenting a likeness of his microscopic subjects.

12 For mentions of drawings of microscopical observations, see for example: Birch, *History of the Royal Society*, vol. 1, 219 (22 April 1663); vol. 1, 231 (29 April 1663); and vol. 1, 234 (6 May 1663).
13 Hooke, *Micrographia*, "The Preface," a2r–v.

Hooke presented his readers with a view of the object that was accepted as accurate. He and his contemporaries were striving to include images with their texts that recorded what they saw and how they saw it; these images were intimately tied to the act of looking. In his overview of the history of scientific illustrations, Brian Ford notes that microscopes present the scientist with problems of interpretation that are not present when looking with the naked eye.[14] As Ford observes, questions arise pertaining to texture and color in particular. These hermeneutic difficulties are relevant particularly at this early moment in the use and dissemination of images based on microscopy. When Samuel Pepys first bought a microscope, he only had Henry Power's microscopical observations to help him try to understand what he saw when looking through the eyepiece. In his *Diary*, he described he and his wife's struggles to "come to find the manner of seeing anything."[15] However, it was only with the publication of Hooke's book a few months later that Pepys became truly enraptured with the wonders that could be revealed with the microscope.[16] How does the user of the microscope separate optical artifacts created by the instrument from the "true form," as Hooke called it? Hooke's early training in the arts and his lifelong interest in prints and printing techniques aided him in the visual decision-making that is involved in looking through a microscope. Accuracy is not something that the viewer perceives in the images, but something that is produced in Hooke's writings about his own process of image making.

In the Preface, Hooke forswore the role of the imagination in his work and more broadly in the reformation of natural philosophy. Instead, he stressed the importance of sensory evidence.

> I hope, they [my endeavors] may be in some measure useful to the main Design of a *reformation* in Philosophy, if it be only by shewing there is not so much requir'd towards it, any strength of *Imagination*, or exactness of *Method*, or depth of *Contemplation*...as a *sincere Hand*, and a *faithful Eye*, to examine, and to record, the things themselves as they appear.[17]

While Hooke claimed his images recorded "the things themselves as they appear," they in fact look quite different from earlier images produced as a

14 Ford, *Images of Science*, 167. For a study of the place of observation in seventeenth-century inquiry see: Daston, "The Empire of Observation," 81–113.

15 *Diary of Samuel Pepys*, vol. 5, 241 (14 August 1664).

16 *Diary of Samuel Pepys*, vol. 6, 18 (21 January 1664/5).

17 Hooke, *Micrographia*, "The Preface," a2v. Italics in the original.

result of looking through a microscope. This chapter shows that the images included in *Micrographia* are the product of more than "a *sincere Hand*, and a *faithful Eye*," and that they are in fact embedded in and indebted to the visual vocabulary used in portrait engraving of the seventeenth century. In both genres the goal of the image is to represent the individual characteristics of the object of study. This was also true of early telescopic images, such as those included in Galileo's *Sidereus nuncius*.[18] It should be noted that the visual vocabulary discussed in this chapter was not unique to portraiture, but rather was widespread throughout the visual culture of seventeenth-century London. In this regard, portraiture represents a test case for comparison with microscopic images based on their shared investments in formal similarity, an object-based focus, and shared conventions in line use.

While it might be assumed that Hooke was depicting an ideal specimen that would help others looking through a microscope to identify what they were seeing, he, in fact, was depicting very particular specimens. For example, in Scheme 31, which shows the back, eyes, and belly of a long-legged spider, figure one shows the spider from above and it has only four legs. (Fig. 2.2) In figure three, which depicts the spider from below, all eight legs are still intact. These images likely show the same spider over the course of Hooke's examinations as it is entirely possible that while manipulating the spider to turn it over some of the legs broke off.[19] This pair of images, like portrait engravings, recorded what his objects of study looked like at a given time. Portrait engravers also sought a balance between an idealized portrayal and a veristic one as the artist had to weigh a sitter's desire to be seen as attractive against the reality of his actual features.[20]

As this example shows, the act of looking is not benign. These views of a spider highlight the fact that the act of representing an object alters it. Indeed, looking and inspecting yields an object that is different than the one that was initially under consideration. While the case of the spider is a very literal example of this, the traces of how looking through the microscope changed the object of study can be seen in other images as well. Hooke's

18 Winkler and Van Helden, "Representing the Heavens," 195–217. Winkler and Van Helden describe Galileo's drawings and the prints made from them as "portraits."

19 I want to thank Steven C. Turner, Curator, Division of Medicine and Science, National Museum of American History for his extraordinary efforts to help me understand what the world looks like through a microscope. While we were studying a flea he had brought in for me to study, I inadvertently pulled off a number of legs while trying to get a better look.

20 Marcia Pointon discusses the balance artists sought in creating likenesses, particularly in the eighteenth century. Pointon, *Hanging the Head*, esp. Chapter III(i), "Likeness and Genre."

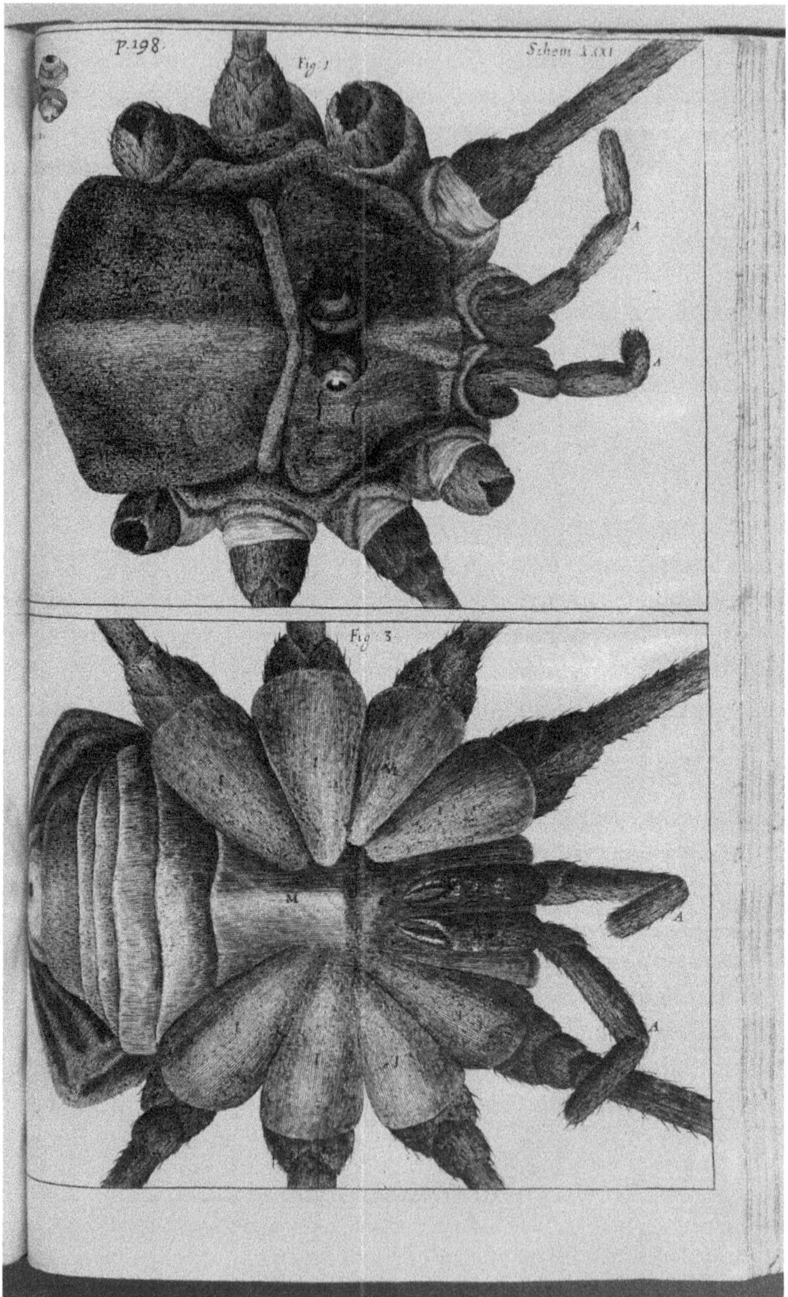

Figure 2.2: Anonymous, *Scheme 31*, Engraving, from Robert Hooke, *Micrographia* (London: Printed by J. Martyn and J. Allestry, 1665). By courtesy of the Department of Special Collections, Memorial Library, University of Wisconsin-Madison, Gift of Daniel and Eleanor Albert, Oversize RE26 O62 H66 M53 1665.

image of deer hair, discussed above, creates the illusion that the viewer of the image is looking through the eyepiece of the microscope at the object of study on the stage. In this image and others like it, the resulting representation presents the object of study as being contained within a perfect circle, therefore limiting the scope of the object to that which can be seen in a single viewing.[21] While the part of the specimen that is cut off by the physical limitations of the instrument is still a part of the object, unlike the spider's legs, all that exists for the viewer is what fit within the scope of the lens. What the viewer is presented then is a view of an object that is not what was put on the stage of the microscope, but rather what resulted from the act of looking through the microscope.

In this chapter, I emphasize the inextricable link between Hooke's involvement with the arts and the development of modern science. As other scholars have noted, Hooke had deep and abiding connections to the London arts community.[22] The most thorough analysis of Hooke's interest in the visual arts is Matthew Hunter's *Wicked Intelligence*. While he is more concerned with broad, theoretical understandings of the act of representing ideas visually, my work considers the pragmatics of making images that pass as accurate. In a similar vein to my work, Jim Bennett has argued that Hooke's connections to instrument makers made him more successful at integrating developments in the mechanical arts into those in natural philosophy.[23] I build on his work to show in addition that Hooke's involvement in the visual arts was key to his work with the microscope. Neither Hooke's dedication to the work being done by the Royal Society nor his active engagement with the arts alone could have led to the production of *Micrographia*. Rather the two together resulted in a book that changed the way the microscopic world was understood.

This chapter begins with an overview of the state of the visual culture of microscopy before Hooke's interventions into it. The second section highlights Hooke's connections to the artistic community and artistic practice in London in the middle of the seventeenth century. The third section examines the

21 One or more figures on eleven of the plates include a view circumscribed by the neat circle of the eyepiece. The images are included on schemes: 2, 3, 5, 9, 10, 11, 12, 14, 15, 20, and 25.

22 See for instance, Hunter, *Wicked Intelligence*. Hunter has also elaborated his argument about Hooke and representation in an article: Hunter, "Theory of the Impression," 167–190. Rob Iliffe also puts a great deal of emphasis on Hooke's connections to artisans, Iliffe, "Material Doubts," 285–318. Lisa Jardine, Margaret 'Espinasse, and Michael Cooper all discuss Hooke's connections to artists and craftsmen. Jardine, *Curious Life of Robert Hooke*, 53–57. 'Espinasse, *Robert Hooke*, 46. Cooper, "Hooke's Career," 1–62. More recently, Felicity Henderson has reassessed Hooke's connections with artists and craftsmen to better understand his role in the Royal Society's acceptance of visual materials. Henderson, "Robert Hooke," 403–418.

23 Bennett, "Hooke's Instruments," 80. See also Dennis, "Graphic Understanding," 309–364.

components of the visual vocabulary of portrait engraving that would have been familiar to Hooke and contrasts them with the vocabulary of the early visual culture of microscopy. Finally, I discuss and show how this vocabulary formed the basis of the regime of accuracy Hooke developed for this "new visible World." Through his understanding and appreciation of the arts of drawing and printing, Hooke created images that changed what accuracy meant for the microscopic world and captivated audiences for generations.

The Visual Culture of Early Microscopy

Although Hooke's *Micrographia* may be the most written-about work of early microscopy, it was by no means the first. The goal of this section is not to rehearse the complicated and contested story of who invented the microscope, but rather to provide a visual overview of the printed images of microscopy that circulated among natural philosophers and natural historians during the first half of the seventeenth century and examine how the terms of the visual vocabulary under consideration were articulated in those early images.[24] That is, how was three-dimensionality conveyed to early viewers of the microscopic world?

With only a few exceptions, all of the images that I will discuss in this section depict insects. As Catherine Wilson has noted, insects were of interest to early users of microscopes in part because of their connections to communicable diseases.[25] In addition, naturalists were concerned with understanding the structure of insects. Wilson also points out that many naturalists felt the need to defend their study of insects, since their detractors dismissed the value of endless hours spent studying creatures often thought to be pests.[26] For the most part then, these early images were representing objects that could be seen by the naked eye with a level of detail not visible without the aid of new optical devices. The challenge then was to convey to viewers the relative proportion of the invisible parts of minute objects, a sense of the depth and weight of these tiny bodies, and the contrasting play of light on their shiny carapaces and fuzzy legs.

A number of attempts were made in the early entries into the visual culture of microscopy to facilitate the viewer's understanding of the relative

24 For an overview of this history see: Ruestow, *Microscope in the Dutch Republic*, esp. chap. 1, "Of Light, Lenses, and Glass Beads," 6–36.

25 Wilson, *Invisible World*, 153–157.

26 Wilson, *Invisible World*, 182–186.

size of the objects of study with the aid of insets. In order to facilitate the viewer's conception of the relative scale of these minute bodies and their understanding of the relative proportion of the parts of the bodies, enlarged sections of the insects were provided. These insets not only provide a more detailed view of a particular part of the insect, but they also provide further information to help viewers unlock the ratio of the parts to the whole. This helped them understand the relative proportion of the parts of the insects to their own lived experience. Since understandings of proportion were based on ratios, it was important to provide the viewer with a frame of reference to help them grasp the level of magnification they were seeing.

The most successful example of this is found in an image embedded in a footnote to a humanist translation of a Roman text. In 1630 Francesco Stelluti (1577–1652), one of the founding members of the Accademia dei Lincei, a group of individuals led by Federico Cesi (1585–1630) who were concerned with the study of nature through close observation, included an image created with the aid of a microscope in a footnote to the word *curculio*, or weevil, in his translation of Persius's satires.[27] (Fig. 2.3) Embedded in the text of the footnote that begins on the previous page, this image combines views of the magnified insect with the unmagnified one to give a sense of the strength of the microscope, which is approximately ten power, and to help the viewer understand the relative scale of the image.[28] The unmagnified weevil is a quarter of an inch long and would therefore have been visible without magnification. What would not have been visible were the articulations of the legs and the details of the eyes and mandibles. The magnified view of the weevil stretches to the full height and width of the plate as the mandibles reach for the top edge and four of the six legs nearly brush the plate mark. The portion of the insect presented in the greatest detail is its mandibles. An inset of them at higher magnification is shown next to the whole insect. This inset is a sketch that shows the jagged, teeth-like structures inside the tip of the mandibles. These three views of the mandibles not only provide a previously unseen level of detail, but also allow the viewer to piece together an understanding of the relative scale and proportion of the weevil. A sense of the proportion of this minute weevil is conveyed not through the skill of the engraver, but rather through the inclusion of additional comparative images.

27 For a thorough study of the Lincei see: Freedberg, *Eye of the Lynx*. Stelluti, *Persio*, 126–127. The plate appears on page 127.

28 The unmagnified weevil is 1/4" long. The magnified whole insect is 2 5/8" long. These measurements would indicate that Stelluti's microscope was approximately x10.

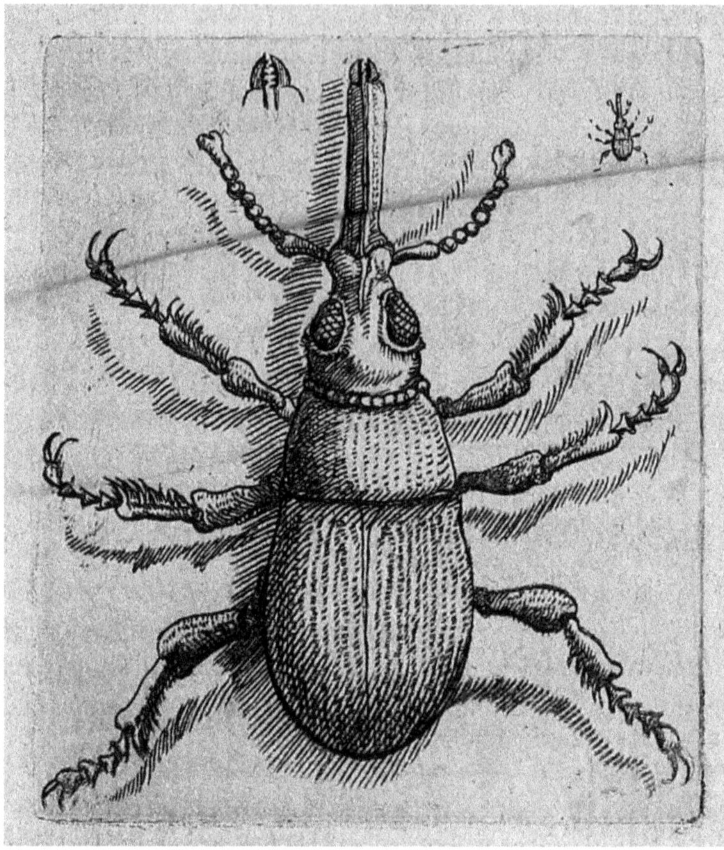

Figure 2.3: Anonymous, *Weevil*, page 127, Engraving, from Francesco Stelluti, *Persio: Tradotto in Verso Sciolto e Dichiarato* (Rome: G. Mascardi, 1630). Dibner Library, Smithsonian Libraries via the Biodiversity Heritage Library.

While this example from 1630 consists of a fairly successful depiction of the relative size of the microscopic parts of a minute object, others were less successful. In 1656 Pierre Borel (ca. 1620–1671) published *Observationum Microcospicarum Centuria* in The Hague. Borel's text consists of one hundred observations on a wide range of objects seen through the microscope, seven of which include illustrations. Insects are the primary subjects illustrated as six of the seven woodcuts represent views of insects; the seventh represents fish scales. One of the images of a whole insect includes a further enlargement of one of the butterfly's antennae.[29] (Fig. 2.4) This antenna most closely resembles a bay branch as the delicate feelers of the antenna are shown as being thick and leaf-like. The full image of the butterfly is just over an

29 This image is embedded in Observation 56, "De Papilionibus." Borel, *Observationum*, 30.

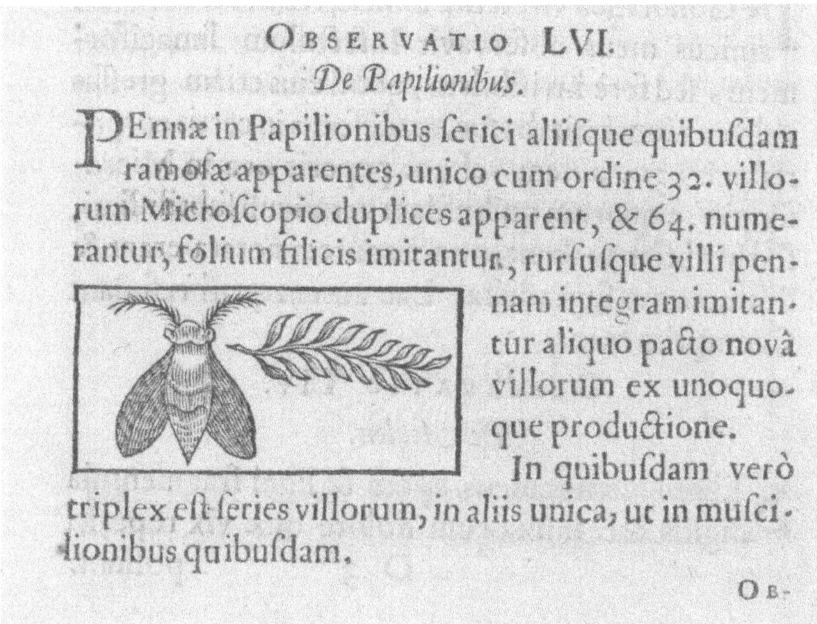

OBSERVATIO LVI.
De Papilionibus.

PEnnæ in Papilionibus ſerici aliiſque quibuſdam ramoſæ apparentes, unico cum ordine 32. villorum Microſcopio duplices apparent, & 64. numerantur, folium filicis imitantur, rurſuſque villi pennam integram imitantur aliquo pacto novâ villorum ex unoquoque productione.

In quibuſdam verò triplex eſt ſeries villorum, in aliis unica, ut in muſcilionibus quibuſdam.

OB-

Figure 2.4: Anonymous, *De Papilionibus (Butterfly)*, page 30, Woodcut, from Pierre Borel, *Observationum Microcospicarum Centuria* (The Hague: Ex typographia Adriani Vlacq, 1655). ETH-Bibliothek Zürich via e-rara.ch.

inch high, which raises questions about whether the whole insect has been magnified or whether it is just the antenna that is enlarged. While it has been estimated that Borel was using a microscope that magnified ten times, the same as Stelluti's, the relative scale of magnified part to the whole insect is not clearly conveyed.[30] Although this image was published twenty-five years after Stelluti's, a consistent method for conveying a sense of proportion in depictions of the microscopic world had not been established.

Hooke resolves these difficulties in *Micrographia* by providing a scale. The first plate of minute objects, Scheme 2, shows highly magnified views of the point of a needle, a printed period, and the edge of a razor. (Fig. 2.5) The text of *Micrographia* starts with an exhortation to begin your microscopic studies with simple objects so that you can learn to discern what you are seeing when you look through the microscope. The concern is that the viewer will lose his way among nature's minute meandering paths and subsequently lose the ability to judge what he sees and might "wander in the *labyrinth* of groundless opinions."[31] The keys to keeping on the path to understanding

30 Fournier, *Fabric of Life*, 14.
31 Hooke, *Micrographia*, 1. Italics in the original.

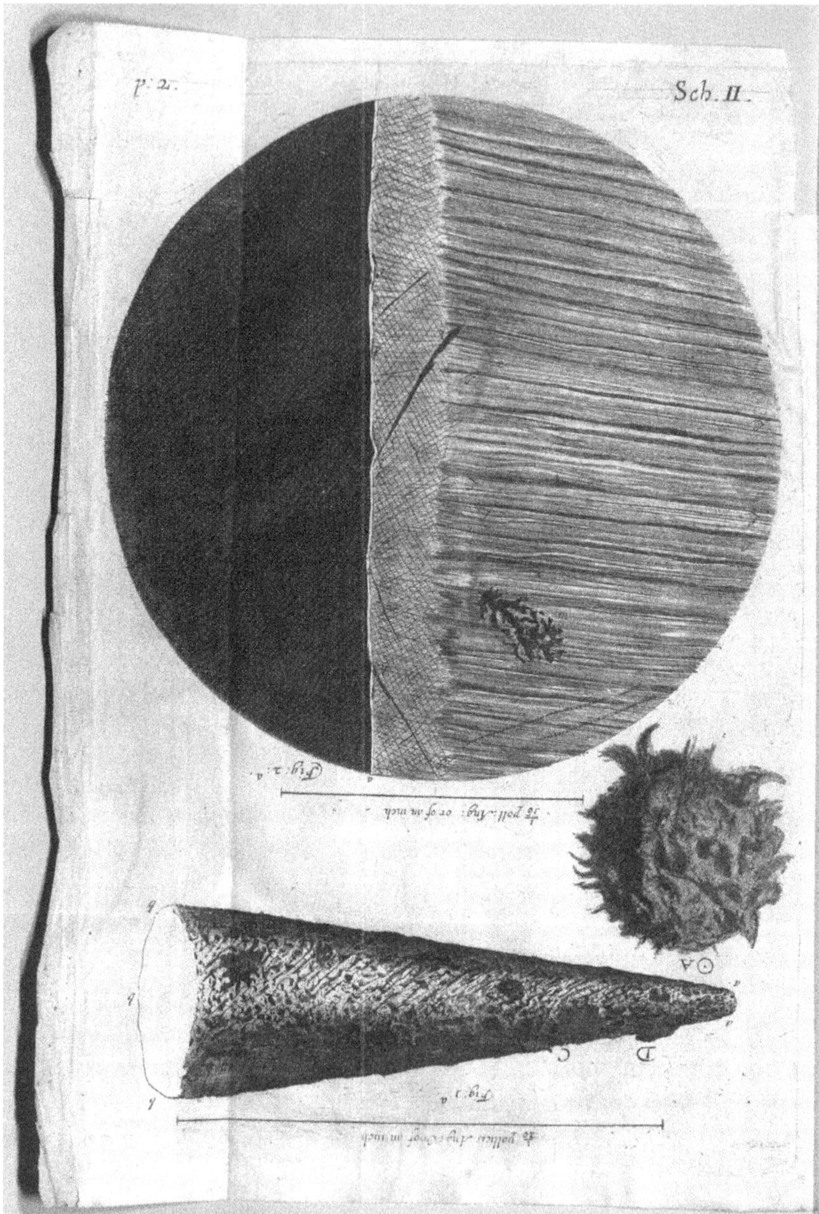

Figure 2.5: Anonymous, *Scheme 2*, Engraving, from Robert Hooke, *Micrographia* (London: Printed by J. Martyn and J. Allestry, 1665). By courtesy of the Department of Special Collections, Memorial Library, University of Wisconsin-Madison, Gift of Daniel and Eleanor Albert, Oversize RE26 O62 H66 M53 1665.

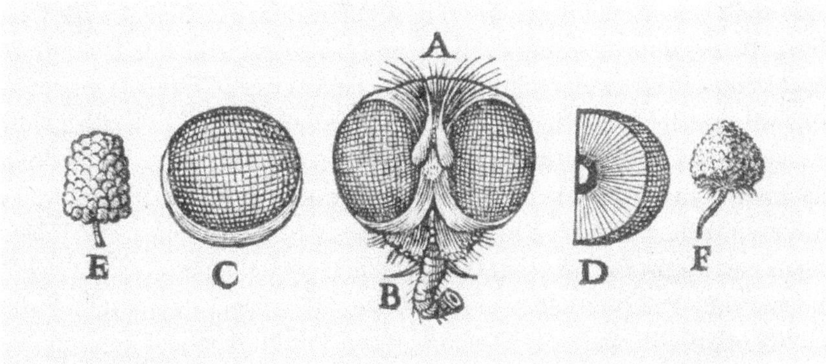

Figure 2.6: Anonymous, *Eye of a Fly*, Woodcut, from Giovanni Battista Hodierna, *L'Occhio della Mosca Discorso Fisico* (Palermo: Per Decio Cirillo, 1644). © British Library Board, (537.c.15.(1.)).

were "*judgment*, that *light*, and *experience*, that *clew*."[32] This combination of ready visual recognition through stark contrast and the introduction to microscopic observation using an uncontested and highly identifiable visual referent sets apart this method. Indeed, the method starts without trying to figure out anything other than how to look at what the observer already knows to be present, and through that knowing process of looking, to begin to discern less readily visible details of the surface or object under magnification. Beginning with familiar objects such as pins and razors helps to ground the observer's studies and Hooke provides further stability in his images by including a linear scale alongside the tip of the pin and the blade of the razor. The relationship of the size of these images to the visible world is given in terms of a known measure, the inch. Even if it is hard to fathom how small 1/26 of an inch is, Hooke has provided his viewers with a clear indication of the relative size of his object of study. He presents his readers the benefit of his own judgment and experience as he attempts to instruct them in the use of the microscope.

There was also a struggle with how to depict a sense of depth in early images of microscopy. For example, Giovanni Hodierna (1597–1660) in his *L'Occhio della Mosca*, published in Palermo in 1644, resorted to analogy to give a sense of the depth and texture of a fly's eye. In this first account of microscopic dissections, the "true form" of the eye is conveyed through its likeness to strawberries and blackberries.[33] The importance of the visual representation of Hodierna's discoveries was made immediately clear as the only image included in this short work appears twice: on the title

32 Hooke, *Micrographia*, 1. Italics in the original.

33 Wilson, *Invisible World*, 76. Fournier, *Fabric of Life*, 26.

page and on page 17 of the text.[34] (Fig. 2.6) The image is therefore the first thing you see upon opening the book and is also embedded in Hodierna's discussion of the results of his dissections. A series of five images spanning the width of the page show a magnified view of a fly's head, a view of a single eye, a dissection of an eye, and two comparative images of a blackberry and a strawberry. The modern estimation is that Hodierna's microscope magnified "no more than twenty to thirty times."[35] While is seems clear from the text that Hodierna had gained an understanding of the anatomy and physiology of insect eyes, his images do not match the quality of his observations.[36] The rough style of this woodcut presents the surface of the fly's eye as a flat grid, without visually presenting a sense of the faceted nature of the eye that he described in the text. The depth and texture of the fly's eye was indicated not through the skill of the woodcutter, but instead through an analogy with the texture of a blackberry and strawberry. However, the depth of the individual components of the fruit is merely suggested in the blackberry by arcing lines piled one onto the other. The dimpled texture of the strawberry was even more vaguely suggested. The images of the berries do not in themselves convey a sense of the depth and texture of a fly's eye, but instead call to mind the actual fruits to aid the viewer in his understanding of the faceted nature of the fly's eye. These images alone were not sufficient; the viewer had to rely on his knowledge of the visible world to understand the invisible. Although Hodierna had gained an understanding of the three-dimensionality of the fly's eye, this depth is not conveyed in the representation of either the eye or the comparative berries.

In looking at how these early entries into the visual culture of microscopy attempt to provide a sense of three-dimensionality for the microscopic world, it becomes clear that authors and the artists working for them were struggling with how to present two-dimensional views of the minute three-dimensional world they saw through the microscope. As much as artist manuals and treatises like Evelyn's stressed the importance of learning to hatch and shade, a vocabulary for how to represent the microscopic world had yet to be established. As a result of his connections to and interest in the London art world, Hooke was able to mobilize an existing visual vocabulary in his depictions of his "new visible World."

34 This is the case in the copy of this work that I consulted in the collection of the British Library (Shelfmark 537.c.15.(1.)).

35 Fournier, *Fabric of Life*, 14.

36 Fournier, *Fabric of Life*, 26–27.

Developing a "sincere Hand and a faithful Eye"

As a visual vocabulary of microscopy emerged, it accompanied a habitus of careful looking that was shaped by an expertise shared among peers, the reinforcement and confirmation of properly discerned features or elements in the objects under scrutiny, and the immediate professional connections that artisans made through a network of practitioners. This section reevaluates Hooke's connection with the arts and artisans in order to better understand the context in which *Micrographia* was produced and highlights the multifaceted nature of Hooke's interests in artistic practice. Using evidence from Birch's *History of the Royal Society* and Hooke's *Diary* refines and nuances a version of Hooke's biography to show how these particular features of his interest in the arts significantly shaped his view of science. While scholars now accept that Hooke understood connections between art and science deeply, my emphasis in this chapter on the particular features of Hooke's habitus, in addition to his propagation of this habitus across his network, contributes to the scholarly knowledge about how Hooke actually brought his training in the visual arts to bear on his scientific endeavors.[37] This section further emphasizes the value placed on Hooke's expertise as a draftsman by the Fellows of the Royal Society.

Hooke's formal involvement in the arts began at a young age, when after the death of his father, in 1648, he was sent to London from his home on the Isle of Wight and apprenticed to Peter Lely (1618–1680).[38] Lely, a Dutch painter, was active in England by 1643 and had completed his obligations to the Painter–Stainers' Company in London in 1647.[39] While it is unclear how long Hooke remained in Lely's studio, he was there at a time when Lely was working to establish himself as artistic heir to Anthony van Dyck and becoming the predominant portrait painter in England.[40] After the Restoration in 1660, his place as painter to Charles II was secure.

Even if Hooke's stay in Lely's studio was brief, it is clear that the two remained in contact after Lely's rise to fame in the Court of Charles II. Lely was among a number of painters who agreed, in 1667, to meet with a committee of Fellows of the Royal Society, which included Hooke, to discuss their

37 Also essential to this research are: Scala, "Index of the Proper Names," 263–329 and Henderson, "Unpublished Material," 129–175.

38 'Espinasse, *Robert Hooke*, 46. John Aubrey in his account of Hooke's life also notes that he "had some instruction in draweing" from Samuel Cooper, who Aubrey described as the "prince of limners of this age." Aubrey, "*Brief Lives*," 410.

39 Dethloff, "Lely, Sir Peter (1618–1680)."

40 Aubrey, *Brief Lives*, 410; Millar and Dethloff, s.v. "Lely, Sir Peter."

"several curiosities and varieties of painting."[41] Furthermore, on December 27, 1675 Hooke recorded in his *Diary* that he went "to Mr. Lelys. Discoursed of helping the sight and of picture box."[42] Hooke, who was the Curator of Experiments at the Royal Society, went to visit Charles II's Court Painter to discuss mechanical aids used in the creation of images. This was not the first mention of the "picture box," or *camera obscura* in Hooke's *Diary*: one had been brought to his house a year earlier.[43] In 1663, Hooke was required, as part of his duties as Curator of Experiments, "To try the casting of a picture on a wall in a light room; and to bespeak a concave glass for it."[44] He performed further experiments with variations on these optical devices dealing with adjustments in scale and righting the inverted image.[45] In 1668, he published a description of his "contrivance to make the picture of anything appear on a wall."[46] Given the central place Svetlana Alpers has shown the *camera obscura* held in discussions of the mechanics of sight and the picturing of the natural world, it should come as no surprise that Hooke and the Fellows of the Royal Society explored the optical workings of the *camera obscura*.[47]

In addition to his interest in the *camera obscura*, Hooke recorded techniques used by artists in the practice of drawing in his *Diary*. He transcribed a lengthy description of "the way to coppy prints…with ease" that was transmitted to him by the same Mr. Kirk who brought the "picture box" to his house a few days later.[48] In the same passage of his *Diary*, Hooke also recorded sharing his technique for drawing ovals.[49] He was not merely interested in theoretical problems related to drawing, but also was a practicing draftsman. He showed many drawings of his experimental findings and instruments of his own invention at meetings of the Royal Society.[50] In addition, he was often commis-

41 Birch, *History of the Royal Society*, vol. 2, 111 (22 August 1667). The group of painters included Lely, Robert Streater, and Samuel Cooper. The prospect of visiting the painters was again raised in December of the same year. Birch, *History of the Royal Society*, vol. 2, 230 (19 December 1667).

42 *Diary of Robert Hooke*, 204.

43 "Mr. Kirk here saw Picture box [e.g. camera obscura] and Raritys." *Diary of Robert Hooke*, 142 (19 January 1674/5).

44 Birch, *History of the Royal Society*, vol. 1, 313 (12 October 1663).

45 See for example, Birch, *History of the Royal Society*, vol. 2, 436 (26 May 1670).

46 Hooke, "A Contrivance," 741–743.

47 Alpers, *Art of Describing*, 27–33. For a collection of essays on artists and the *camera obscura* see, *Inside the Camera Obscura*, edited by Wolfgang Lefévre.

48 *Diary of Robert Hooke*, 140 (6 January 1674/5).

49 *Diary of Robert Hooke*, 140 (6 January 1674/5).

50 For drawings of instruments, see for example: Birch, *History of the Royal Society*, vol. 1, 287 (July 29, 1663); vol. 1, 295 (19 August 1663) and Henderson, "Unpublished Material," 143 (11 July 1672). For a detailed discussion of a series of drawings Hooke did of fossils see: Kusukawa, "Drawings of Fossils," 123–138.

sioned by the Fellows to produce drawings of specimens that were brought into meetings and of experiments performed by Fellows outside of the scheduled meeting time, so that others could see the results.[51] Furthermore, Hooke served as an arbiter of the accuracy and usefulness of drawings that others sent in to the Royal Society with their reports of observations.[52] Hooke's training and interest in drawing meant that he both contributed drawings to the Society's meetings and also determined which best served the Society's ends.

Hooke's practical interest in drawing extended to the books he read. In July of 1675 he borrowed "a book of limning" from William Faithorne, who sold prints and books by others as well as himself.[53] Two weeks later, he recorded copying the book he had borrowed from Faithorne.[54] While it is unclear exactly which book Hooke borrowed from Faithorne as there were numerous books on the arts of drawing and limning available in London during the second half of the seventeenth century, we do know that Hooke owned a copy of the 1666 edition of Thomas Jenner's *A Book of Drawing, Limning, Washing or Colouring of Maps and Prints*.[55] Furthermore, Matthew Hunter has attributed a drawing at the Tate Britain to Hooke that appears to be based on another mid-seventeenth-century drawing manual, Alexander Browne's *Ars Pictoria*.[56] The result of this combined interest shapes how we understand Hooke's relationship with images. Hooke was concerned both with mechanical aids for drawing and manuals intended for young draftsmen. The range of his inquires show that he was not simply interested in aspects of drawing that were closely linked to questions of concern to natural philosophers, i.e. the *camera obscura*, but also with the more practical aspects of learning to draw. These interests, in turn, informed his own practice of drawing.

Hooke's curiosity extended beyond drawing techniques to include techniques for printing. In an undated manuscript, Hooke transcribed the first chapter of Abraham Bosse's *Traicté des manières de graveur en taille douce*, which details how to make the varnish that is used to coat the

51 While many of these drawings have been lost, a number of them do still exist in the Archives of the Royal Society including that of the heart stones, Register Book Original, vol. 2, 66 (RBO/2i/66). Hooke was asked to draw the stones that were taken out of the Earl of Belcarres's heart; Birch, *History of the Royal Society*, vol. 1, 292 (5 August 1663). This drawing, and the copying of it, are discussed in Fransen, Reinhart, and Kusukawa, "Copying Images," 263. He also presented drawings based on Hevelius's and Cassini's reports of comet sighting. Henderson, "Unpublished Material," 142 (19 June 1672).

52 See for instance, Birch, *History of the Royal Society*, vol. 2, 31–32 (19 April 1665).

53 *Diary of Robert Hooke*, 169 (12 July 1675).

54 *Diary of Robert Hooke*, 170 (24 July 1675).

55 Hooke's copy is now in the collection of the British Library (Shelfmark 536.l.21.(3.)).

56 Hunter, "Hooke's Figurations," 254.

copper plate when making an etching.[57] He recorded in his *Diary* numerous techniques he invented for improving printmaking as well as recording reports of improvements for the art of printing suggested by others.[58] He even offered advice on the subject to his friend John Ogilby (1600–1676), a London publisher and geographer. Hooke wrote that he "shewd him [Ogilby] the way of Letters for marking his map and also the way of shadowing."[59] This passage implies that Hooke was giving Ogilby tips for preparing drawings for printing as the maps in question were probably part of one of Ogilby's atlases.[60] Of particular importance to my study of the images in *Micrographia* is that Hooke was giving advice on "the way of shadowing" indicating that he believed he had a superior grasp on how to indicate the fall of light within a two-dimensional image. Hooke believed he knew the best ways of preparing drawings in order for them to be rendered in print.

In addition to sharing advice on printing techniques with gentlemen, Hooke was on friendly terms with a group of London-based printmakers and purchased prints and books on art and architecture from a number of booksellers around London. Wenceslas Hollar is perhaps the most recognizable name among the list of Hooke's associates.[61] Hooke went to Hollar's on numerous occasions and recorded buying several his prints. He also bought works by Jacques Callot and Albrecht Dürer among others.[62] Along with buying prints, Hooke recorded having books by Giorgio Vasari, a biography of Peter Paul Rubens, and a variety of books on the history of printing.[63] These are just a

57 Archives of the Royal Society, Classified Paper 24 (CLP/24/88).

58 On June 29, 1674 Hooke recorded that he had "Invented the way of printing with the common press pictures made with Pinns. An Invention of Great use." *Diary of Robert Hooke*, 109–110. For notes on others' innovations see for instance: *Diary of Robert Hooke*, 17 (21 December 1672); 19 (3 January 1672/3); 24 (27 January 1672/3); 57 (26 August 1673); 65 (14 October 1673); 95 (3 April 1674); and 143 (22 January 1674/5).

59 *Diary of Robert Hooke*, 65 (14 October 1673).

60 For a discussion of Hooke's connections with Ogilby and his atlas projects see: Taylor, "Robert Hooke," 529–540.

61 Others in the list include William Faithorne, David Loggan, Mr. Davys the Younger, Wat Dole (probably Walter Dole, Faithorne's apprentice), and Mr. Lamb. These names appear throughout Hooke's *Diary*: Hollar: 54 (9 August 1673), 55 (16 August 1673), 108 (16 June 1674), 136 (15 December 1674), 249 (14 September 1676), 379 (2 October 1678); Davys: 112 (11 July 1674), 115 (31 July 1674), 116 (7 August 1674), 118 (22 August 1674), 125 (6 October 1674); Dole: 95 (3 April 1674 and 4 April 1674); Lamb: 85 (6 February 1673/4) and 134 (7 December 1674, 9 December 1674, and 10 December 1674).

62 For works by Callot see *Diary of Robert Hooke*, 10 (17 October 1672) and 136 (15 December 1674); for works by Dürer see 170 (23 July 1675).

63 "brought from Arundel Library 3 volumes of Vasari of the Lives of Painters, Cesari Ripa of Iconologia, Rubens Life etc." *Diary of Robert Hooke*, 38 (9 April 1673). Hooke records owning copies of John Evelyn's *Sculptura* (90 (7 March 1673/4)), Abraham Bosse's *Traicté des manières*

sampling of the interactions Hooke had with artists and booksellers regarding prints and books on the arts. Hooke's *Diary* is filled with references to meetings that attest to his accumulation of prints and books and the auction of his books after his death contained over ninety books related to the visual arts.[64]

All of these connections indicate that he was deeply embedded in the visual culture of seventeenth-century England. Whether in discussions with painters or other Fellows of the Royal Society, Hooke's *Diary* shows he was intrigued by different image-making strategies and conversed with both makers and consumers of images about technologies of vision and visualization. Further, Birch's *History of the Royal Society* is filled with references to Hooke's "draughts" and "schemes." Beginning with his early education in Lely's studio and continuing throughout his life, Hooke was involved with the production and consumption of images. These connections to the arts must be understood within the context of his advocacy of "a *sincere Hand*, and a *faithful Eye*" as being essential to "a *reformation* in Philosophy."

Making "a Plain Representation"

This section outlines how portrait engravings provided the visual lexicon that Hooke made use of in preparing his drawings to be engraved. In describing his method for creating the "draughts" that were used as the basis for the engraved plates in *Micrographia*, part of which served as the epigraph to this chapter, Hooke made it clear that he did not immediately draw what he saw upon looking through his microscope because of the difficulties of discerning what he saw:

> ...in making them [the drafts], I indeavoured (as far as I was able) first to discover the true appearance, and next to make a plain representation of it. This I mention the rather, because of these kind of Objects there is much more difficulty to discover the true shape, then of those visible to the naked eye, the same Object seeming quite differing, in one position to the Light, from what it really is, and may be discover'd in another. And therefore I never began to make any draughts before by many

de graveur en taille douce (291 (17 May 1677)), and recorded having "Bought of Pits, *History of printing* 1sh" (121 (9 September 1674)).

64 Leona Rostenberg has analyzed and published this catalogue. Rostenberg, *Library of Robert Hooke*, 133. More recently, Will Poole, Felicity Henderson and Yelda Nasifoglu has created a detailed and searchable site for studying Hooke's library: https://hookesbooks.yeldanasifoglu.com/.

> examinations in several lights, and in several positions to those lights, I had discover'd the true form.[65]

While others have noted the use of similar language to travel narratives in this passage, what is striking in the context of understanding the production of accuracy is his repeated emphasis on discovering truth.[66] The "truth" of these objects was not readily apparent upon first glance. Hooke had to make a series of decisions about what exactly it was that he was seeing while preparing his drawings. These decisions were based on past experience of the visual world and required Hooke to make judgments in order "to distinguish between a *prominency* and a *depression*, between a *shadow* and a *black stain*, or a *reflection* and a *whiteness in the colour*."[67] These terms are representative of the visual vocabulary with which Hooke was engaged. In this passage, Hooke was particularly concerned with how to interpret and represent the depth and contrast he observed in "these kind of Objects" and his training as a draftsman informed his creation of "a plain representation."

Looking at an engraving from the period provides a better sense of how Hooke deployed a visual vocabulary of engraving. A portrait of Charles II engraved by William Faithorne from the 1650s serves as an example of the portraits that were circulating in London at the time that Hooke was preparing the drawings for *Micrographia*.[68] (Fig. 2.7) By studying Faithorne's likeness of the king, we gain an understanding of the visual resources available to Hooke in the process of determining what he saw through the microscope.

Faithorne's rendering of the king's hair gives us an opportunity to think about how engravers conveyed how "to distinguish between a *prominency* and a *depression*." The waves of Charles's hair are communicated through several means: first, sinuous lines convey the overall contours of his hair; second, the depressions are made darker through a layering of hatched lines, by which the darkest areas are understood as the low points of the waves; and finally, the prominences of the waves of hair are indicated as the white of the paper shows through larger gaps in the engraved lines. The relatively sparse lines on the crests of the waves give the illusion of light hitting these high points. This combination of curving lines and varied density of lines

65 Hooke, *Micrographia*, "The Preface," f2v.
66 See for example: Campbell, *Wonder & Science*.
67 Hooke, *Micrographia*, "The Preface," f2v. The emphasis is in the original.
68 In all likelihood, the "J.O." who penned the verses below the image of Charles is John Ogilby, mentioned earlier as someone with whom Hooke discussed drawing techniques.

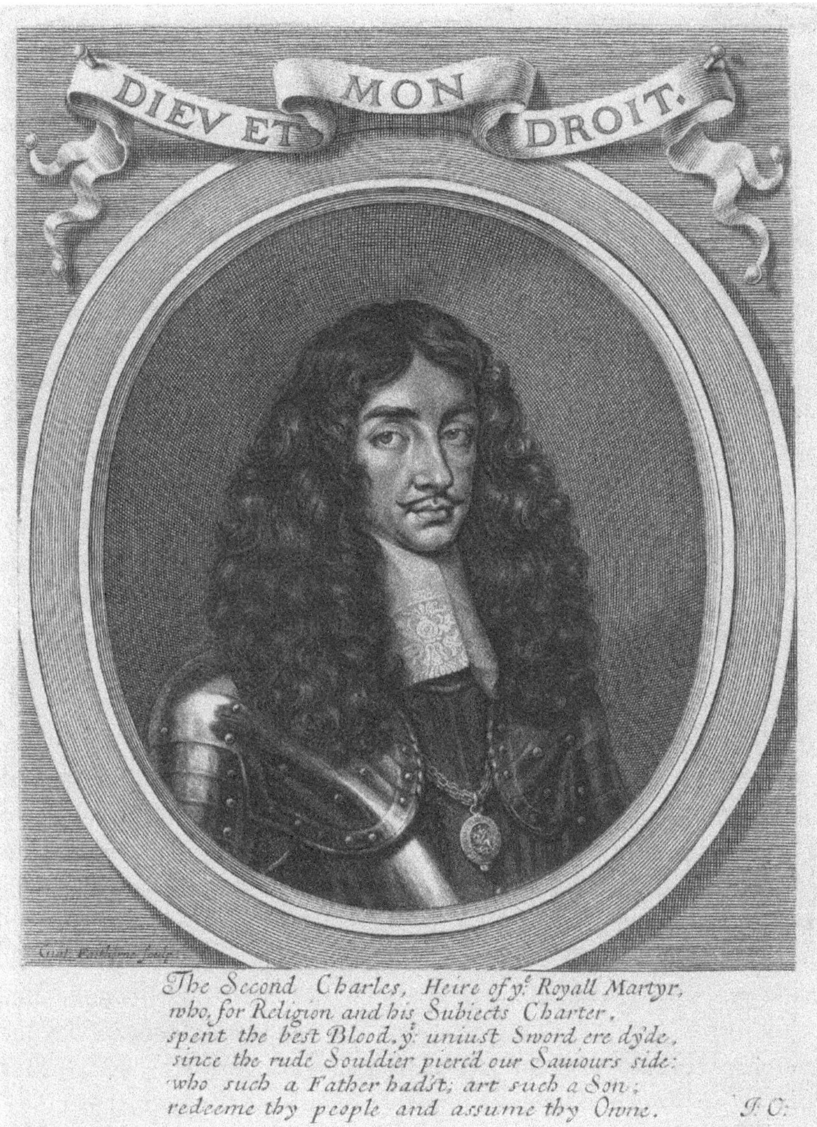

Figure 2.7: William Faithorne, *Portrait of Charles II*, Engraving, 1650s. Courtesy National Gallery of Art, Washington, D.C., Ailsa Mellon Bruce Fund, 2006.4.2.

creates an illusionistic representation of thick, wavy hair with definite prominences and depressions. The three-dimensionality of Charles's curls is depicted in two dimensions through variations in the shape and density of the engraved lines and is in sharp contrast to Hodierna's fly eye and berries.

Comparing Charles's armor with his collar indicates how "a *reflection* and a *whiteness in the colour*" are distinguished according to a visual vocabulary of engraving. In both cases the whiteness of the paper shows through, but it is quite clear in looking at this image that his armor is reflecting the light and his collar is white. What I am interested in is *how* this is made so clear to a viewer. An obvious answer is that it is unreasonable to think that his armor has a white stripe in it. However, there is more to it than that. The alternation of black and white, dark and light, in the area surrounding the highlight on his right shoulder visually adds up to a reflection. The area behind the hot spot on his shoulder is heavily shaded with a series of overlapping lines that taper down the shoulder and hug the bright spot. In addition, the area around the hot spot gets gradually brighter at the edges of the reflection. Increasing brightness is indicated by a progressive decrease in the density of lines; the lines become both thinner and fewer. The irregularity of the shape of the white spot on his shoulder also adds to our sense that we are looking at a reflection. This is not a round white marking on the armor; instead, the light plays off the curve of the metal as it bends around the form of his shoulder. The bright white of the paper combined with the dark shading on the upper part of the shoulder give a clear indication that we are looking at a reflection, not a "whiteness in the colour." The contrast of light and dark gives a clear sense of the play of light on a three-dimensional object.

In contrast with the glaring white of the reflection, the whiteness of the collar is represented more evenly. The plain white section of it is indicated with regularly spaced lines that curve slightly and get thicker where there is a dimple in the fabric about halfway between the patterned section and his neck. The more shadowed side of the collar is also indicated with very even lines that are now crossed with diagonals indicating the shadow cast on the collar by the King's chin. Whiteness is not indicated through unadulterated paper. Instead, lighter lines with more space between them convey the whiteness of Charles's collar. While reflections are shown through striking contrasts of dark and light, whiteness is expressed through an evenness and regularity of line.

Even Hooke's closest contemporary, Henry Power (ca.1626–1668), struggled with finding a visual vocabulary for depicting the subtleties of the contrast of light and dark of the microscopic world in his *Experimental Philosophy*, published in London in 1664. Power's book is composed of three sections of "new experiments: microscopical, mercurial, magnetical."[69] There are three woodcut illustrations included in the first section of the book treating

69 Power, *Experimental Philosophy*, title-page.

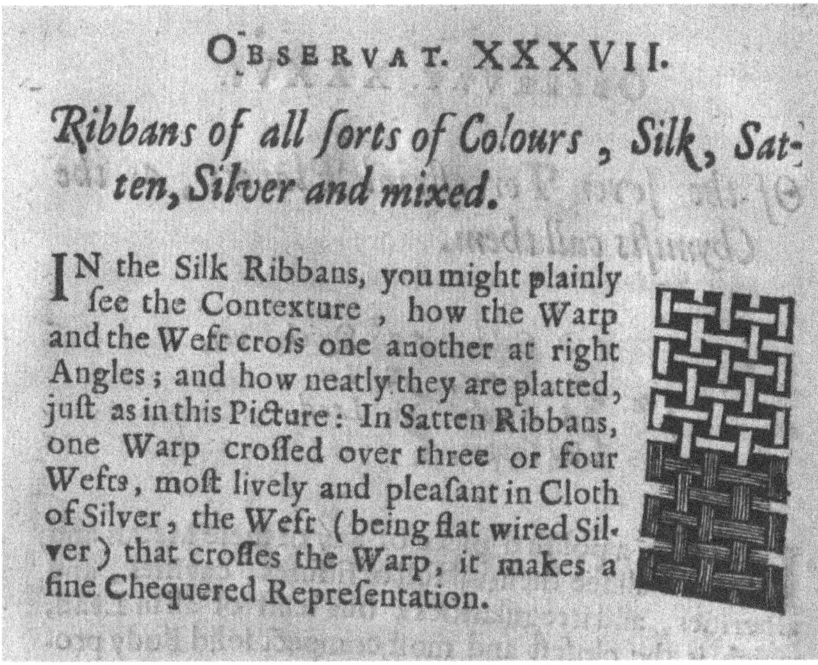
OBSERVAT. XXXVII.

Ribbans of all ſorts of Colours, Silk, Satten, Silver and mixed.

IN the Silk Ribbans, you might plainly ſee the Contexture, how the Warp and the Weft croſs one another at right Angles; and how neatly they are platted, juſt as in this Picture: In Satten Ribbans, one Warp croſſed over three or four Wefts, moſt lively and pleaſant in Cloth of Silver, the Weft (being flat wired Silver) that croſſes the Warp, it makes a fine Chequered Repreſentation.

Figure 2.8: Anonymous, *Ribban of all sorts of Colours*, page 46, Woodcut, from Henry Power, *Experimental Philosophy, in Three Books* (London: Printed by T. Roycroft for John Martin and James Allestry, 1664). By courtesy of the Department of Special Collections, Memorial Library, University of Wisconsin-Madison, Thordarson T 2332.

"experiments microscopical." Power chose to illustrate his observations of the eyes of spiders, a piece of ribbon, and poppy seeds with small woodcuts. The illustration of the warp and weft of a silk ribbon only vaguely indicated how the light hits the different fibers. (Fig. 2.8) The image gave a clear sense of the fibers meeting at right angles, upon which Power commented in the text.[70] However, there is no nuance in the shading of the image, neither in the stark upper section, which gives no indication of the texture of the fibers, nor in the more subtle lower section. At the points of intersection between the warp and the weft, the space around the crossed fibers is solid black. There is no sense of the gradual darkening of the appearance of one strand as it goes under another. An area that would appear shaded when you looked at it is simply rendered in black. The small woodcuts in Power's book do not convey a sense of the play of light off of the silk ribbon that one would expect. Although Power's book and Hooke's shared the same publisher and they were both Fellows of the Royal Society, they do not

70 Power, *Experimental Philosophy*, 46.

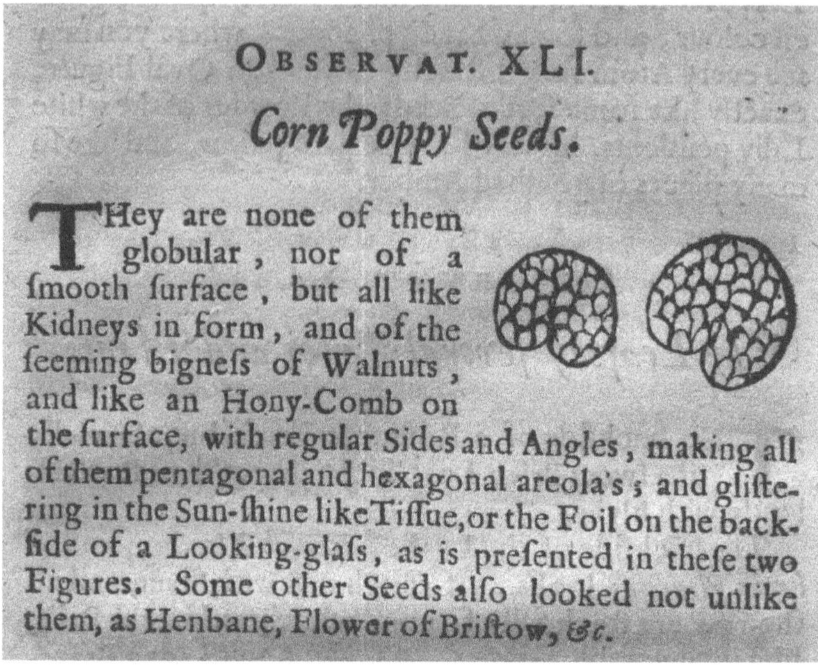
OBSERVAT. XLI.

Corn Poppy Seeds.

THey are none of them globular, nor of a ſmooth ſurface, but all like Kidneys in form, and of the ſeeming bigneſs of Walnuts, and like an Hony-Comb on the ſurface, with regular Sides and Angles, making all of them pentagonal and hexagonal areola's; and gliſtering in the Sun-ſhine like Tiſſue, or the Foil on the back-ſide of a Looking-glaſs, as is preſented in theſe two Figures. Some other Seeds alſo looked not unlike them, as Henbane, Flower of Briſtow, &c.

Figure 2.9: Anonymous, *Corn Poppy Seeds*, page 49, Woodcut, from Henry Power, *Experimental Philosophy, in Three Books* (London: Printed by T. Roycroft for John Martin and James Allestry, 1664). By courtesy of the Department of Special Collections, Memorial Library, University of Wisconsin-Madison, Thordarson T 2332.

share the same vocabulary for depicting the microscopic world. Power's illustrations are much closer to Hodierna's in terms of media and style than they are to Hooke's.

These examples show how engraved lines in portraits were combined to indicate differences in depth and in light and dark and how this vocabulary was not used by Hooke's predecessors. The play of light on the undulating waves of Charles's curls and the bright reflection off his armor are indicated through variations in the shape and density of lines. Portraits of a wide range of individuals created by many different artists in the seventeenth century used this same vocabulary, and as we will see, so did Hooke.

Engraving "the True Form"

While the large-format plates depicting the minute details of insects are perhaps the best known in *Micrographia*, I want to focus my discussion of the importance of the visual vocabulary of portrait engraving on a humbler topic: the poppy seed. As we saw earlier, this was one of the three objects that Henry Power illustrated in the section on microscopy in his *Experimental Philosophy*. (Fig. 2.9 & 2.10) The contrast between Power's and Hooke's views of poppy seeds attests to the lengths Hooke went to depict the "true form" of his objects of study and to create an illusion of accuracy. Although published within a year of one another by the same printers, these two representations give very different senses of the three-dimensionality of the microscopic world. It would be easy to dismiss the differences between these two images as being solely tied to media. That is, one could argue that because Power's illustration is a small woodcut, it is limited by the constraints of the medium. However, this argument would discount the existence of highly detailed and nuanced woodcuts that were published by artists such as Albrecht Dürer.[71] Rather, it was Hooke's involvement with the arts and his awareness of the techniques of draftsmen and engravers that yielded such striking images.

While their most noticeable difference is scale, I am less interested in the relative size of these images and more interested in how these images convey a sense of the microscopic texture of the poppy seed and the depth of the craters on the surface.[72] Power's small woodcut image makes clear that the surface of a poppy seed is not as smooth as meets the naked eye. What is unclear from this image is whether the surface of the seed is made up of a series of overlapping scales like those found on a fish, or whether there is some other explanation. Power's text provides more clues as to what he saw:

> ...like an Hony–Comb on the surface, with regular Sides and Angles, making all of them pentagonal and hexagonal areola's and glistering in the Sun-shine like Tissue, or the Foil on the backside of a Looking-glass, as is presented in these two Figures.[73]

The lines delineating the texture of the poppy seeds, however, do not meet at crisp angles that would suggest honeycombs. In addition, the uniform blackness of the lines gives no sense of the glistening of sunshine or the shimmer of mirrored glass. Although Power seemed to believe that his image helped to clarify his textual description ("as is presented in these two Figures"), the woodcuts fell short of the standards of description established

71 See for example his Apocalypse series: Mende, *Albrecht Dürer*, no. 97–112 and Strauss, *Sixteenth Century German Artists*, vol. 10, no. 60–75.

72 Power's larger poppy seed is 3/4 by 3/4" whereas one of Hooke's is 3 by 2 3/8".

73 Power, *Experimental Philosophy*, 49.

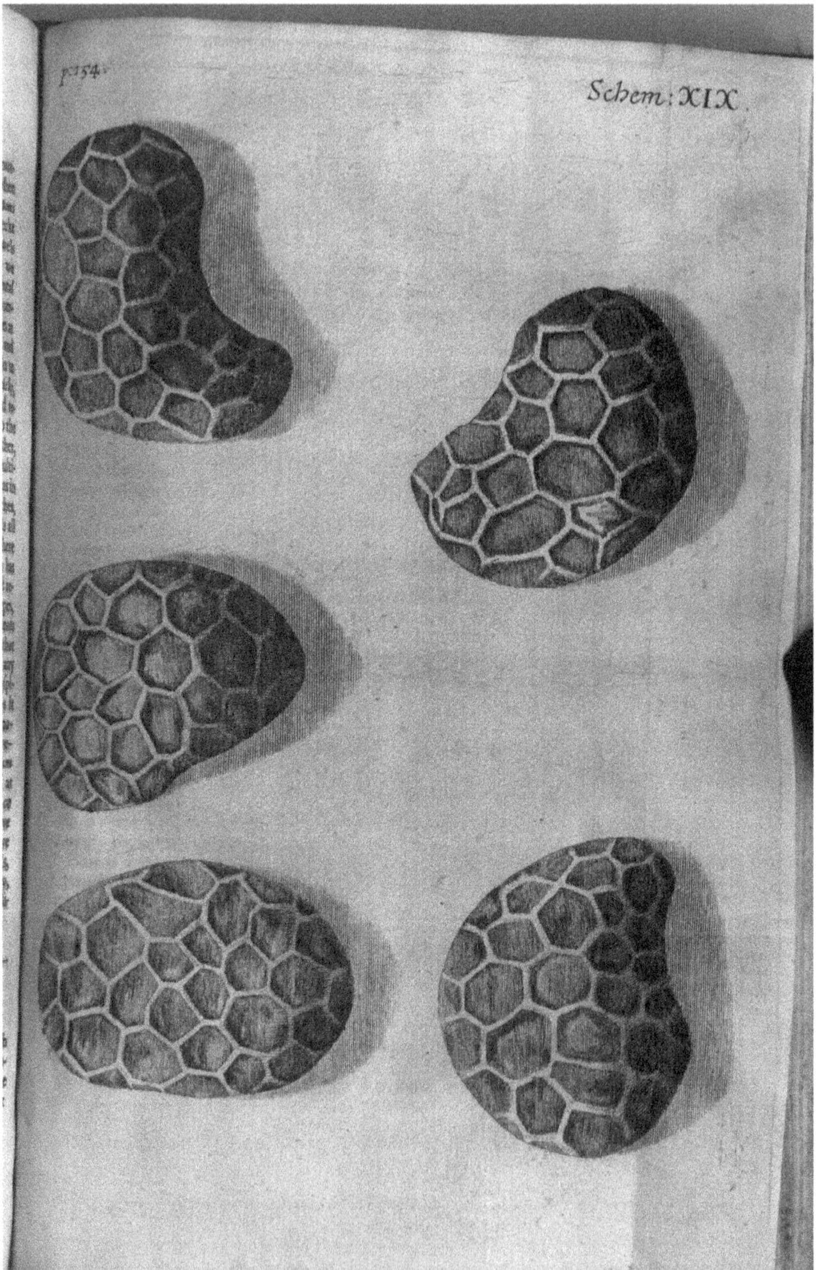

Figure 2.10: Anonymous, *Scheme 19*, Engraving, from Robert Hooke, *Micrographia* (London: Printed by J. Martyn and J. Allestry, 1665). By courtesy of the Department of Special Collections, Memorial Library, University of Wisconsin-Madison, Gift of Daniel and Eleanor Albert, Oversize RE26 O62 H66 M53 1665.

by the text. As we saw in Hodierna's *L'Occhio della Mosca*, this discrepancy between text and image was common. The images in Hooke's *Micrographia* close this gap by applying the visual vocabulary of portrait engraving to the representation of objects under microscopic investigation.

If we turn to Hooke's plate of poppy seeds, we can see the crisp angles of the honeycomb formations and the bright reflection of light glistening on the uneven surface of the seeds. Like Power, Hooke remarked on the grid-like structure of the surface of the seeds in his text, and, again like Power, he used analogies to describe what he saw:

> ...curiously Honey-comb'd all over with a very pretty variety of Net-work, or a small kind of imbosment of very orderly rais'd ridges, the surface of them looking not unlike the inside of a Beev's [Beef's] stomack.[74]

Hooke's images, however, rise to the level of his textual description. In the images of each of these five seeds it is possible "to distinguish between a *prominency* and a *depression*, between a *shadow* and a *black stain*, or a *reflection* and a *whiteness in the colour*."[75]

As we saw in the hair in the portrait of Charles II, in this plate variations in light and dark indicated through differences in the density of the lines convey a sense of the depth of the "imbosments" that Hooke saw. The ridges of each cell cast distinct shadows into the basin as the light falls onto each seed from the left. For instance, looking at the seed in the upper right hand side of the plate, it is clear that the large oblong basin on this seed is particularly deep, as the lines are so thick as to become almost black and they fade out into the lit portion that is outside of the shadow of the indentation. The same visual techniques are used in the engravings in *Micrographia* as in Faithorne's portrait.

Whereas in Power's image any variation in the black lines that delineated the honeycombs was the result of the uneven way that the paper took up the ink, in Hooke's plate the seeds do seem to be "glistering in the Sun-shine like Tissue, or the Foil on the backside of a Looking-glass."[76] This illusion is created through the same vocabulary that made the King's armor glisten. The bright white of the paper shines through on the tops of the "rais'd ridges" and the sides of the ridges are heavily shaded, like the back of Charles's

74 Hooke, *Micrographia*, 155. I want to thank Sean Clark, Professor of Agriculture and Natural Resources, Berea College, for pointing out to me that the reticulum of a bovine digestive system has a honeycomb-like surface and is sometimes referred to as honeycomb tripe.

75 Hooke, *Micrographia*, "The Preface," f2v.

76 Power, *Experimental Philosophy*, 49.

shoulder. In addition, in the same seed in the upper right, one of the ridges is so directly in the light as to reflect the light cast on it. We know from Hooke's description that the seeds are "of a dark brownish red colour."[77] In addition, we know from looking at Charles's armor that a bright white section of paper showing through does not indicate a white spot. Hooke's text combined with our understanding of the visual vocabulary of engraving clearly indicate that we are seeing a reflection and not "a *whiteness in the colour*."

While this plate showing five poppy seeds is just one of the thirty-eight included in *Micrographia*, it shows clearly how different Hooke's images were from earlier ones, and the comparison with Faithorne's portrait of Charles II helps us to understand *how* that difference is accomplished. It would be easy to dismiss this difference as dependent on the engravers and not on Hooke, but that would be underestimating Hooke's commitment to the visual component of his book. As we have seen, he spent a great deal of time looking at his objects of study before drawing them. In addition, he writes in the Preface that "…in divers of them the Gravers have pretty well follow'd my directions and draught…"[78] While this may sound like slight praise, Hooke was well known in the period as being someone who was willing to openly express his displeasure should his ire be stoked, so this comment should be read as an endorsement for the engravings.[79]

There are very few extant drawings from Hooke's hand related to the production of *Micrographia*, but there are many drawings connected to other projects.[80] A leaf of drawings in a manuscript at the British Library includes a few drawings of insects that have been attributed to Hooke.[81] One of the figures, of a stick-like bug, shows how the visual vocabulary of

77 Hooke, *Micrographia*, 155.

78 Hooke, *Micrographia*, "The Preface," f2v.

79 See for instance his conflict with Newton. Margaret 'Espinasse began her biography of Hooke with an account of how Hooke's death in 1703 was the condition upon which it was possible for Newton to become President of the Royal Society. 'Espinasse, *Robert Hooke*, 1. Lisa Jardine, similarly began her account of Hooke's life with his dispute with Newton. Jardine, *Curious Life*, 1–19.

80 The extant drawings of snowflakes (MS/215/7) and frozen urine (Cl.P. 20/5) are essentially renderings of two-dimensional objects and are therefore less useful for understanding Hooke's ability to translate the three-dimensional world into a two-dimensional rendering. For a reproduction of one of the few drawings related to *Micrographia* see: Harwood, "Rhetoric and Graphics," plates 9a and 9b. Volume 20 of the Classified Papers in the Archives of the Royal Society is comprised of papers written by Hooke. Many of these papers include sketches by Hooke that clarify and augment the arguments he made in the text. For the repeated copying of Hooke's drawings of snowflakes, see: Fransen and Reinhart, "Practice of Copying," 211–216.

81 The drawing is folio 113v in Add. MS 57495 at the British Library. Janice Neri has attributed this drawing to Hooke. Neri, "Some Early Drawings," 41–47; Neri, "Between Observation and Image," 95–99; and Neri, *Insect and the Image*, 123–131.

portrait engraving was incorporated into Hooke's drawing style. (Fig. 2.11) We can see clearly in this drawing how the light reflected off the insect's back. The pen strokes along the ridge of the back become sparser as they approach the high point and finally on the highest section, the paper shows through as we saw in Faithorne's depiction of Charles's armor. Even though this insect does not appear in *Micrographia*, it provides concrete evidence that Hooke's drawings as well as the printed images in the book owe a clear and substantial debt to the visual vocabulary of portrait engraving.

Other drawings of Hooke's in the Archives of the Royal Society also attest to his grasp of how to translate three dimensions into two. For example, in a paper on the refraction of ice written in 1663, Hooke includes a drawing of the set up for his experiment.[82] (Fig. 2.12) The reader is directed to this figure to gain a better understanding of how to recreate the experiments themselves. The figure gives a clear sense of the "cylindrical glasse" that Hooke used as the lines shading the front of the vessel curve around the right-hand edge and become thicker and darker towards the bottom as the vessel curves down towards the small feet holding it up off the surface of the table. Hooke indicates the three-dimensionality of the vessel using the same vocabulary that Faithorne used to indicate the curve of Charles's armor around his shoulder. These two examples of how Hooke used the visual vocabulary of engraving in his drawings add to our understanding of how Hooke's involvement and interest in the arts influenced how he saw the "new visible World" he discovered with his microscope.

Although Hooke's *Micrographia* was only published one year after Power's *Experimental Philosophy* and John Martyn and James Allestry, Printers to the Royal Society, published them both, the images included in their books are markedly different. As this chapter has shown, this difference is due in part to Hooke's connections to the artistic community in London and his interest in printing and the printed image. Further, the effect of accuracy was produced through visual references to other images. The plates in *Micrographia* shocked and fascinated viewers in the seventeenth century. Samuel Pepys recorded staying up until "two o'clock in my chamber reading of Mr. Hooke's Microscopicall Observations, the most ingenious book that ever I read in my life."[83] The book also circulated throughout Europe and was the basis of the illustrations in many other books on microscopy. When it was reviewed in the *Journal des Sçavans*, ten of the eleven pages of the issue

82 This drawing appears in an essay "Of the refraction of ice," which was read at a meeting of the Royal Society on 11 February 1663. Archives of the Royal Society, CLP/20/11.
83 *Diary of Samuel Pepys*, vol. 6, 18 (21 January 1664/5).

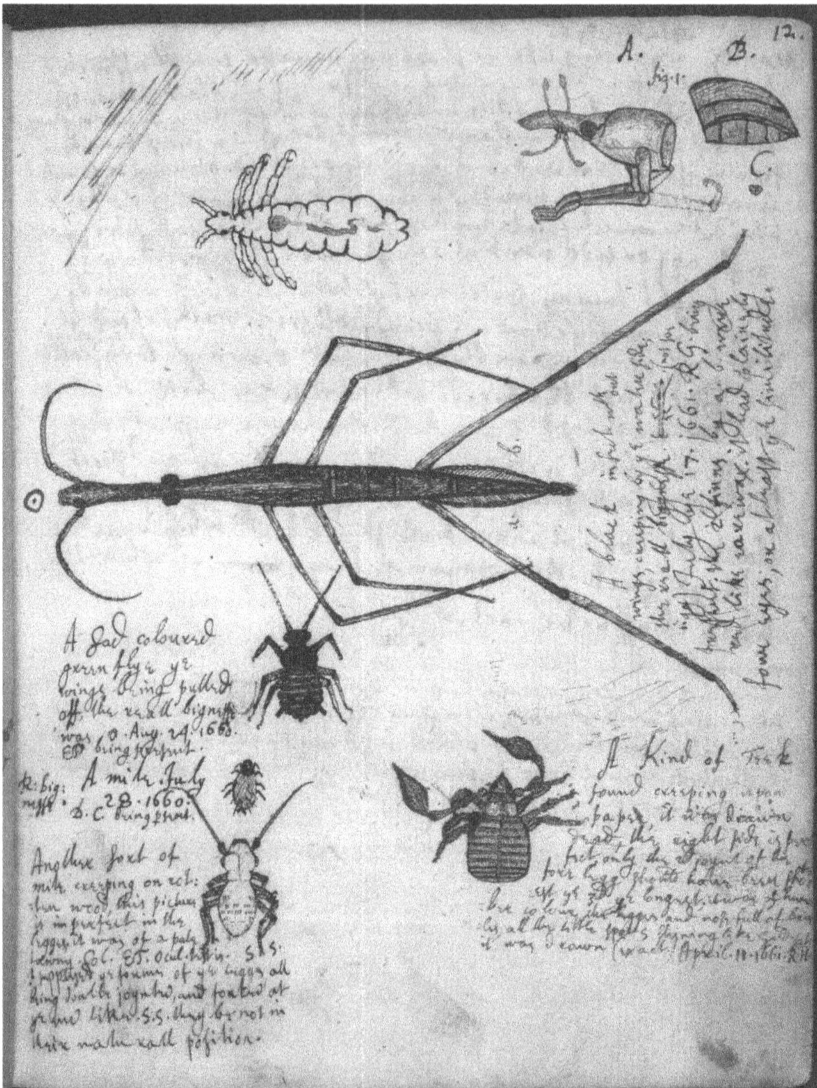

Figure 2.11: Robert Hooke, *Drawing of an Insect*, Ink on paper, from John Covel, *Natural History Notebook and Commonplace Book*, 1660–1713. © British Library Board, Add. MSS 57495.

were given over to a discussion of Hooke's book, and the plates of the louse and mould were copied to show the French readers some of the microscopic wonders Hooke revealed.[84] Even those who disagreed with Hooke copied his images in their books, such as a text by a Jesuit priest published in 1691

84 "Micrographia," 491–501.

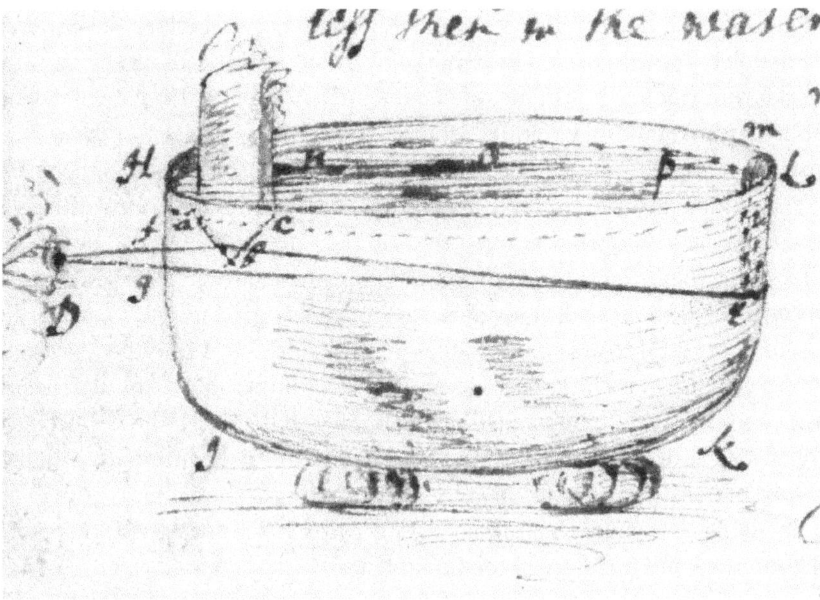

Figure 2.12: Robert Hooke, *Drawing of Experimental Setup*, Ink on paper, from "Of the Refraction of Ice," 1662. © The Royal Society, CLP/20/11.

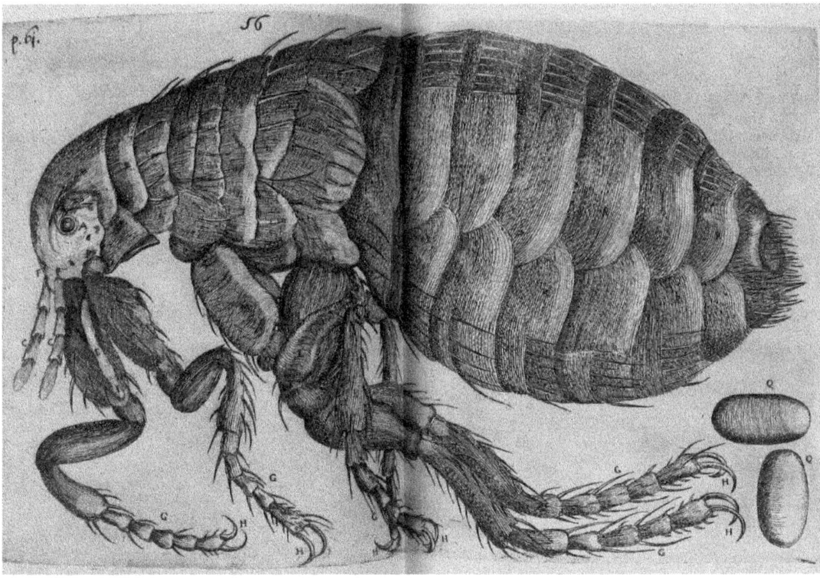

Figure 2.13: Anonymous, *A Flea*, Engraving, from Filippo Buonanni, *Micrographia Curiosa sive Rerum Minutissimarum Observations* (Rome: Typis Dominici Antonij Herculis, 1691). Dibner Library, Smithsonian Libraries via the Biodiversity Heritage Library.

arguing for an Aristotelian view of spontaneous generation.[85] (Fig. 2.13) Although Filippo Buonanni held markedly different views about the natural world from Hooke, he still included copies of a number of Hooke's plates in his own book. Johann Griendel, who set out to create a new *Micrographia*, also copied Hooke plates, but without the same level of visual skill or detail.[86] (Fig. 2.14) In addition, Hooke's plates continued to be printed long after his death. In 1745 Henry Baker reprinted Hooke's plates along with new text explaining them.[87] A further edition was published in 1780.[88] Hooke's images became the standard point of reference for depicting the microscopic world.

For over one undred years Hooke's remarkable plates provided viewers with a glimpse of the microscopic world that vividly called to mind a three-dimensional world using a visual vocabulary that clearly and coherently translated three dimensions into two. The use of this familiar vocabulary helped to construct an illusion of the act of looking at the visible world. That is, by using the visual vocabulary of engraving Hooke's images presented the microscopic world in a recognizable idiom that created a sense of accuracy because the images appeared to represent a three-dimensional world that could only be seen with the aid of "artificial organs." I have highlighted crucial aspects of this idiom through an attention to Hooke's use and understanding of how to render and transmit line, angle, light, reflection, regular and irregular surfaces, as well as the physical features of magnification under the microscope such as relative scale, and the preparation of samples for study. Well beyond the Royal Society's Fellows, those who adopted Hooke's idiom similarly engaged in the careful creation and use of a familiar idiom to yield images that were accepted as representations of what Hooke called the "new visible world." These images were deeply embedded in the act of looking at a particular object. Hooke's discovery of their true form was informed by his own training and interest in the visual arts. Hooke filled the space between looking at an object and picturing one with his knowledge of artistic practice. In so doing, he developed a regime of accuracy that allowed others to accept the usefulness of his images for the communication of new knowledge-.

85 Buonanni, *Micrographia Curiosa*.

86 Griendel, *Micrographia Nova*. The book was presented by Hooke at a meeting of the Royal Society and described as: "giving the figures of several insects, seeds, &c. many the same with those in Mr. Hooke's *Micrograpphia*, but much worse designed." Birch, *History of the Royal Society*, vol. 4, 541 (8 June 1687).

87 Hooke, *Micrographia Restaurata*.

88 Hooke, *Microscopic Observations*.

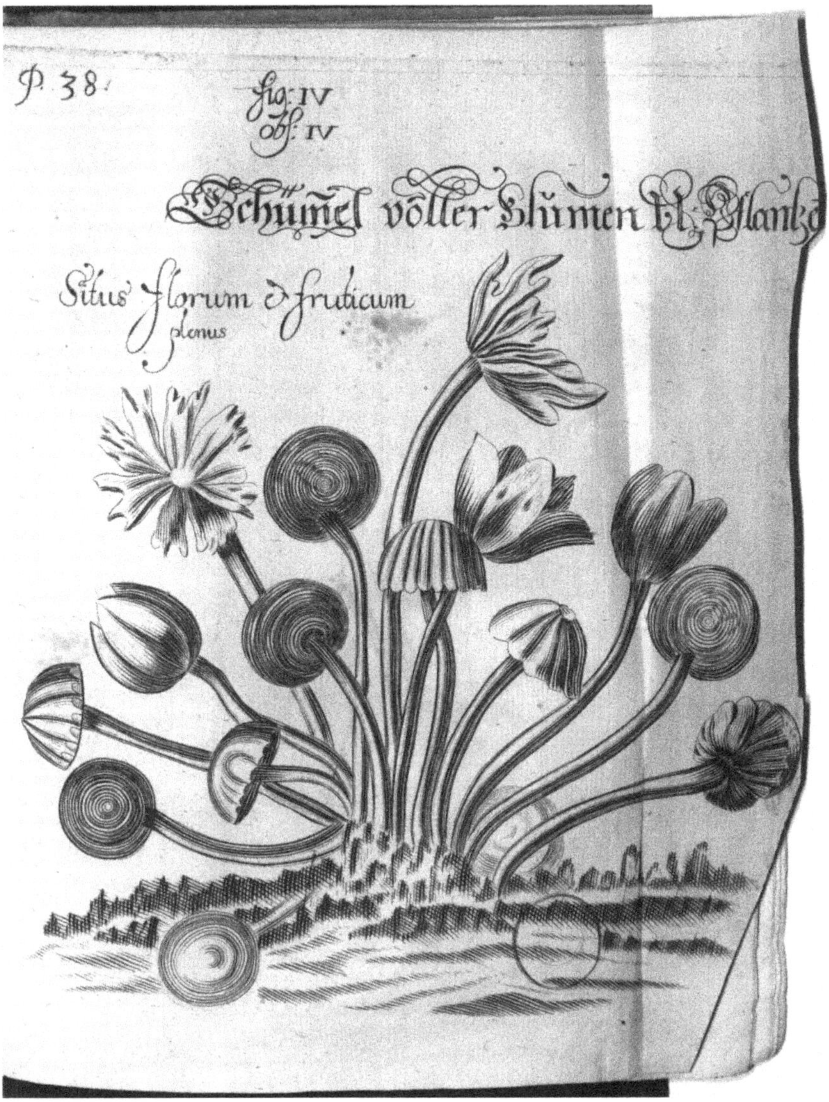

Figure 2.14: Anonymous, *Mold Spores, Figure 4*, Engraving, from Johann Franz Griendel, *Micrographia Nova, Sive, Nova et Curiosa Variorum Minutorum Corporum* (Nuremberg: Sumptibus Johannis Ziegeri, 1687). By courtesy of the Department of Special Collections, Memorial Library, University of Wisconsin-Madison, acquired through the Robert Stauffer Fund, CA 16870.

Bibliography

I. Manuscript Sources

London, Archives of the Royal Society
CLP/20. Classified Papers. Hooke Papers.
CLP/24. Classified Papers. Collins, Oldenburg, Hooke.
MS215. Copies of Papers read to the Society about 1662–1664.
RBO/2i. Register Book Original.

London, British Library
Add. MS 57495. John Covel, D.D., (b.1638, d.1722): natural history notebook and commonplace book; 1660–1713, n.d. Contains many pen and ink drawings of plants, insects, fish, etc. with detailed descriptive text.

II. Printed Primary Sources

Aubrey, John. *"Brief Lives," Chiefly of Contemporaries, Set Down by John Aubrey, between the Years 1669 and 1696*. Edited by Andrew Clark. Oxford: Clarendon Press, 1898.
Birch, Thomas. *The History of the Royal Society of London*. London: Printed for A. Millar, 1756.
A Book of Drawing, Limning, Washing or Colouring of Maps and Prints: And the Art of Painting, with the Names and Mixtures of Colours used by the Picture-Drawers. Or, The Young-mans Time well Spent. London: Printed by M. Simmons, for Thomas Jenner, 1666.
Borel, Pierre. *Observationum Microcospicarum Centuria*. The Hague: Ex typographia Adriani Vlacq, 1655.
Buonanni, Filippo. *Micrographia Curiosa sive Rerum Minutissimarum Observations*. Rome: Typis Dominici Antonij Herculis, 1691.
Griendel, Johann Franz. *Micrographia Nova, Sive, Nova et Curiosa Variorum Minutorum Corporum*. Nuremberg: Sumptibus Johannis Ziegeri, 1687.
Hodierna, Giovanni Battista. *L'Occhio della Mosca Discorso Fisico*. Palermo: Per Decio Cirillo, 1644.
Hooke, Robert. "A Contrivance to Make the Picture of Anything Appear on a Wall, Cup-board, or within a Picture-frame, &c. in the Midst of a Light Room in the Day-time; or in the Night-time in any Room that is Enlightned with a Considerable Number of Candles; Devised and Communicated by the Ingenious Mr. Hook." *Philosophical Transactions* 3, no. 38 (17 August 1668): 741–743.

—. *The Diary of Robert Hooke, M.A., M.D., F.R.S., 1672–1680, Transcribed from the Original in the Possession of the Corporation of the City of London (Guildhall Library)*. Edited by Henry W. Robinson and Walter Adams. London: Taylor & Francis, 1935.

—. *Microscopic Observations or, Dr. Hooke's Wonderful Discoveries by the Microscope, Illustrated by Thirty-Three Copper-Plates, Curiously Engraved: Whereby the Most Valuable Particulars in That Celebrated Author's Micrographia Are Brought Together*. London: Printed for Robert Wilkinson, 1780.

—. *Micrographia, or, Some Physiological Descriptions of Minute Bodies Made*. London: Printed by J. Martyn and J. Allestry, 1665.

—. *Micrographia Restaurata or, the Copper-Plates of Dr. Hooke's Wonderful Discoveries by the Microscope, Reprinted and Fully Explained: Whereby the Most Valuable Particulars in That Celebrated Author's Micrographia Are Brought Together in a Narrow Compass*. Edited by Henry Baker. London: Printed for and sold by John Bowles, 1745.

"Micrographia, or Some Physiological descriptions of minute Bodes made by magnifying glasses, with observations and inquires thereupon, by R. Hooke fellow the Royal Society. In fol. London." *Journal des Sçavans*, 2 (20 December 1666): 491–501.

Pepys, Samuel. *The Diary of Samuel Pepys: A New and Complete Transcription*. Edited by Robert Latham and William Matthews. Berkeley: University of California Press, 1970.

Power, Henry. *Experimental Philosophy, in Three Books: Containing New Experiments Microscopical, Mercurial, Magnetical*. London: Printed by T. Roycroft for John Martin and James Allestry, 1664.

Stelluti, Francesco. *Persio: Tradotto in Verso Sciolto e Dichiarato*. Rome: G. Mascardi, 1630.

III. Secondary Sources

Alpers, Svetlana. *The Art of Describing: Dutch Art in the Seventeenth Century*. Chicago: University of Chicago Press, 1983.

Bennett, Jim. "Hooke's Instruments," in *London's Leonardo: The Life and Work of Robert Hooke*, 63–104. Oxford: Oxford University Press, 2003.

Bloom, Terrie F. "Borrowed Perceptions: Harriot's Maps of the Moon." *Journal of the History of Astronomy* 9 (1978): 117–122.

Campbell, Mary Baine. *Wonder & Science: Imagining Worlds in Early Modern Europe*. Ithaca: Cornell University Press, 1999.

Cooper, Michael. "Hooke's Career," in *London's Leonardo: The Life and Work of Robert Hooke*, 1–62. Oxford: Oxford University Press, 2003.

Daston, Lorraine. "The Empire of Observation, 1600–1800," in *Histories of Scientific Observation*. Edited by Lorraine Daston and Elizabeth Lunbeck. Chicago: University of Chicago Press, 2011, 81–113.

Dennis, Michael Aaron. "Graphic Understanding: Instruments and Interpretation in Robert Hooke's Micrographia." *Science in Context* 3, no. 2 (1989): 309–64.

Dethloff, Diana. "Lely, Sir Peter (1618–1680)." In *Oxford Dictionary of National Biography*. Oxford: Oxford University Press, 2004.

Doherty, Meghan C. "Discovering the 'True Form:' Hooke's Micrographia and the Visual Vocabulary of Engraved Portraits." *Notes and Records of the Royal Society of London* 66, no. 2 (2012): 211–234.

Edgerton, Samuel Y. *The Heritage of Giotto's Geometry: Art and Science on the Eve of the Scientific Revolution*. Ithaca: Cornell University Press, 1991.

'Espinasse, Margaret. *Robert Hooke*. London: William Heinemann, 1956.

Fransen, Sietske and Katherine M. Reinhart. "The Practice of Copying in Making Knowledge in Early Modern Europe: An Introduction." *Word & Image* 35, no. 3 (2019): 211–222.

Fransen, Sietske, Katherine M. Reinhart, and Sachiko Kusuwaka. "Copying Images in the Archives of the Early Royal Society." *Word & Image* 35, no. 3 (2019): 256–276.

Ford, Brian J. *Images of Science: A History of Scientific Illustration*. London: The British Library, 1992.

Fournier, Marian. *The Fabric of Life: Microscopy in the Seventeenth Century*. Baltimore: Johns Hopkins University Press, 1996.

Freedberg, David. *The Eye of the Lynx: Galileo, His Friends, and the Beginnings of Modern Natural History*. Chicago: University of Chicago Press, 2002.

Harwood, John T. "Rhetoric and Graphics in Micrographia." In *Robert Hooke: New Studies*, edited by Michael Cyril William Hunter and Simon Schaffer, 119–47. Woodbridge, Suffolk: The Boydell Press, 1989.

Henderson, Felicity. "Robert Hooke and the Visual World of the Early Royal Society." *Perspectives on Science* 27, no. 3 (May-June 2019): 395–434.

—. "Unpublished Material from the Memorandum Book of Robert Hooke, Guildhall Library MS 1758." *Notes and Records of the Royal Society* 61 (2007): 129–175.

Hunter, Matthew C. "Hooke's Figurations: A Figural Drawing Attributed to Robert Hooke." *Notes and Records of the Royal Society* 64, no. 3 (2010): 251–260.

—. "The Theory of the Impression According to Hooke." in *Printed Images in Early Modern Britain: Essays in Interpretation*. Edited by Michael Hunter. Ashgate: Farnham, Surrey, 2010, 167–190.

—. *Wicked Intelligence: Visual Art and the Science of Experiment in Restoration London*. Chicago: University of Chicago Press, 2013.

Iliffe, Rob. "Material Doubts: Hooke, Artisan Culture and the Exchange of Information in 1670s London." *The British Journal for the History of Science* 28, no. 3 (1995): 285–318.

Jardine, Lisa. *The Curious Life of Robert Hooke: The Man Who Measured London.* London: HarperCollins, 2003.

Kusukawa, Sachiko. "Drawings of Fossils by Robert Hooke and Richard Waller." *Notes and Records of the Royal Society* 67, no. 2 (2013): 123–138.

Lefévre, Wolfgang, ed. *Inside the Camera Obscura: Optics and Art under the Spell of the Projected Image.* Preprint 333. Berlin: Max Planck Institute for the History of Science, 2007.

Mende, Matthias. *Albrecht Dürer: Zum Leben, Zum Holzschnittwerk, Chronologisches Verzeichnis der Holzschnitte, Hinweise zum aktuellen Schrifttum, Zeittafel, Kondordanz.* Munich: Edition Tomas, 1976.

Millar, Oliver and Diana Dethloff. *Grove Dictionary of Art Online,* s.v. "Lely, Sir Peter." Oxford: Oxford Art Online, 2003.

Neri, Janice. "Between Observation and Image: Representations of Insects in Robert Hooke's *Micrographia.*" In *The Art of Natural History: Illustrated Treatises and Botanical Paintings, 1400–1850*, edited by Amy R. W. Meyers and Therese O'Malley, 83–107. Washington: National Gallery of Art, 2008.

—. "Some Early Drawings by Robert Hooke." *Archives of Natural History* 32, no. 1 (2005): 41–47.

—. *The Insect and the Image: Visualizing Nature in Early Modern Europe, 1500–1700.* Minneapolis: University of Minnesota Press, 2011.

Pointon, Marcia. *Hanging the Head: Portraiture and Social Formation in Eighteenth-Century England.* New Haven: Yale University Press, 1993.

Poole, William, Felicity Henderson, and Yelda Nasifoglu. "Robert Hooke's Books." Accessed October 29, 2020. https://hookesbooks.yeldanasifoglu.com/.

Rostenberg, Leona. *The Library of Robert Hooke: The Scientific Book Trade of Restoration England.* Santa Monica, CA: Modoc Press, Inc., 1989.

Ruestow, Edward G. *The Microscope in the Dutch Republic: The Shaping of Discovery.* Cambridge: Cambridge University Press, 1996.

Scala, Gail Ewald. "An Index of the Proper Names in Thomas Birch 'A History of the Royal Society' (London, 1756–1757)." *Notes and Records of the Royal Society* 28, no. 2 (1974): 263–329.

Strauss, Walter L. *Sixteenth Century German Artists: Albrecht Dürer.* Vol. 10. *The Illustrated Bartsch.* New York: Abaris Books, 1980.

Taylor, E.G.R. "Robert Hooke and the Cartographical Projects of the Late Seventeenth Century (1666–1696)." *The Geographical Journal* 90, no. 6 (1937): 529–40.

Turner, Gerard L'E. "The Impact of Hooke's *Micrographia* and its influence on microscopy." In *Robert Hooke and the English Renaissance*, edited by Paul Kent and Allan Chapman, 124–145. Herefordshire: Gracewing, 2005.

Wilson, Catherine. *The Invisible World: Early Modern Philosophy and the Invention of the Microscope*. Princeton, N.J.: Princeton University Press, 1995.

Winkler, Mary G., and Albert van Helden. "Representing the Heavens: Galileo and Visual Astronomy." *Isis* 83, no. 2 (1992): 195–217.

3. "Nearly Resembling the Live Birds": Collecting and Collating for the Reformation of Natural History

Abstract

This chapter examines the plates in Francis Willughby's *Ornithology* and unearths the sources used by the engravers to produce plates that nearly resembled the life, to use John Ray's phrase. While Ray asserted he was not repeating textual error put forth by his predecessors, such as Gessner and Aldrovandi, he used their illustrations as the basis for his own as well as drawings of live and dead birds that he collected. By uncovering the original sources for the illustrations, this chapter argues that greater value was placed on recognizable, printed images than on drawings collected by the author. This conclusion leads to a larger argument about the perceived truth-content of printed natural historical images in the seventeenth century.

Keywords: Francis Willughby, John Ray, Ornithology, Collecting, *Ad vivum*

In the Preface to the English edition of Willughby's *Ornithology*, John Ray described at length the visual resources available to him while he was compiling Francis Willughby's research and adding his own further investigations to it. He had known Willughby since their time at Cambridge, and Ray took on the task of compiling and supplementing his observations after his untimely death at the age of thirty-seven in 1672.[1] Ray provided details about the manuscript collections that Willughby bought while in Strasbourg and Nuremberg as well as drawings he commissioned upon

1 Raven, *John Ray*, 163–167. A more recent edited volume has reassessed the working relationship between Ray and Willughby: Birkhead, ed., *Virtuoso by Nature*.

Doherty, M.C., *Engraving Accuracy in Early Modern England: Visual Communication at the Early Royal Society*. Taylor & Francis Group, 2022.
DOI 10.5117/9789463721066_CH03

his return to England.[2] He also listed some of the printed sources that he used: Aldrovandi, Clusius, Olina, Marggraf, Piso, and Bondt.[3] In addition to having access to earlier printed ornithological texts and drawings, Ray consulted live and dried specimens of some of the birds he was studying.[4] After discussing the difficulties he had communicating with the engravers from where he was working in the country at Willughby's family home, he states: "Notwithstanding the Figures, such as they are, take them all together, they are the best and truest, that is, most like the live Birds, of any hitherto engraven in Brass" (Preface, n.p.). He also described the images as "nearly resembling the live birds" on the title page. This chapter explores what it meant in seventeenth-century London to be "most like the live Birds" and to nearly resemble life by excavating the nuanced process involved in selecting the visual source material used by the engravers for the 373 images of birds included in the eighty plates that accompanied the English edition of Willughby's *Ornithology* (London, 1678).[5] (Fig. 3.1) Prints, drawings, texts, and observations were cross-referenced to create the illusion of accuracy through Ray's repeated enactments of judgment.[6]

By unearthing the complex interrelationships among the images published in Willughby's *Ornithology*, previously printed images, manuscript images, and live and dried specimens, this chapter concludes that "the best and truest" images were not the result of direct observation. That is, life-like and "taken from the life" were not equivalent terms. As others have pointed out, the phrase "from the life" is a vexed designation as it was used

2 Willughby and Ray, *Ornithology of Francis Willughby*, Preface, n.p. Subsequent citations of this book will be made parenthetically in the text. For a detailed description of the sources the images, see my analysis in: Tim Birkhead, et al., "Willughby's Ornithology," 268–304.

3 Aldrovandi, *Ornithologiae*; Clusius, *Exoticorum Libri Decem*; Olina, *Uccelliera*; Piso, *Historia Naturalis Brasiliae*; Piso, *De Indiae*; and Bondt, *Historiæ Naturalis*.

4 As Nathan Fils has shown, Ray also used paintings by Francis Barlow as a source. Fils, "Francis Barlow," 263–300.

5 I am focusing on the English edition because another plate of birds and two plates on trapping were added to the seventy-seven plates included in the Latin edition of 1676. The count of 373 includes each bird depicted on the seventy-eight plates of specimens and each scene on the two plates of trapping techniques. Willughby and Ray, *Ornithologiae libri tres*.

6 In addition to the many sources discussed in this chapter that *were* used in the production of the book, Willughby and Ray amassed a huge collection of drawings that were not used for the images that ended up in the book. Many of these remained with the Willughby family papers and are now in the archives at the University of Nottingham. Mi LM 24: Collection of water-colour drawings and engravings of birds; n.d. mid-seventeenth-early eighteenth century. For a discussion of this collection see: Grindle, "'No other sign'," 15–22.

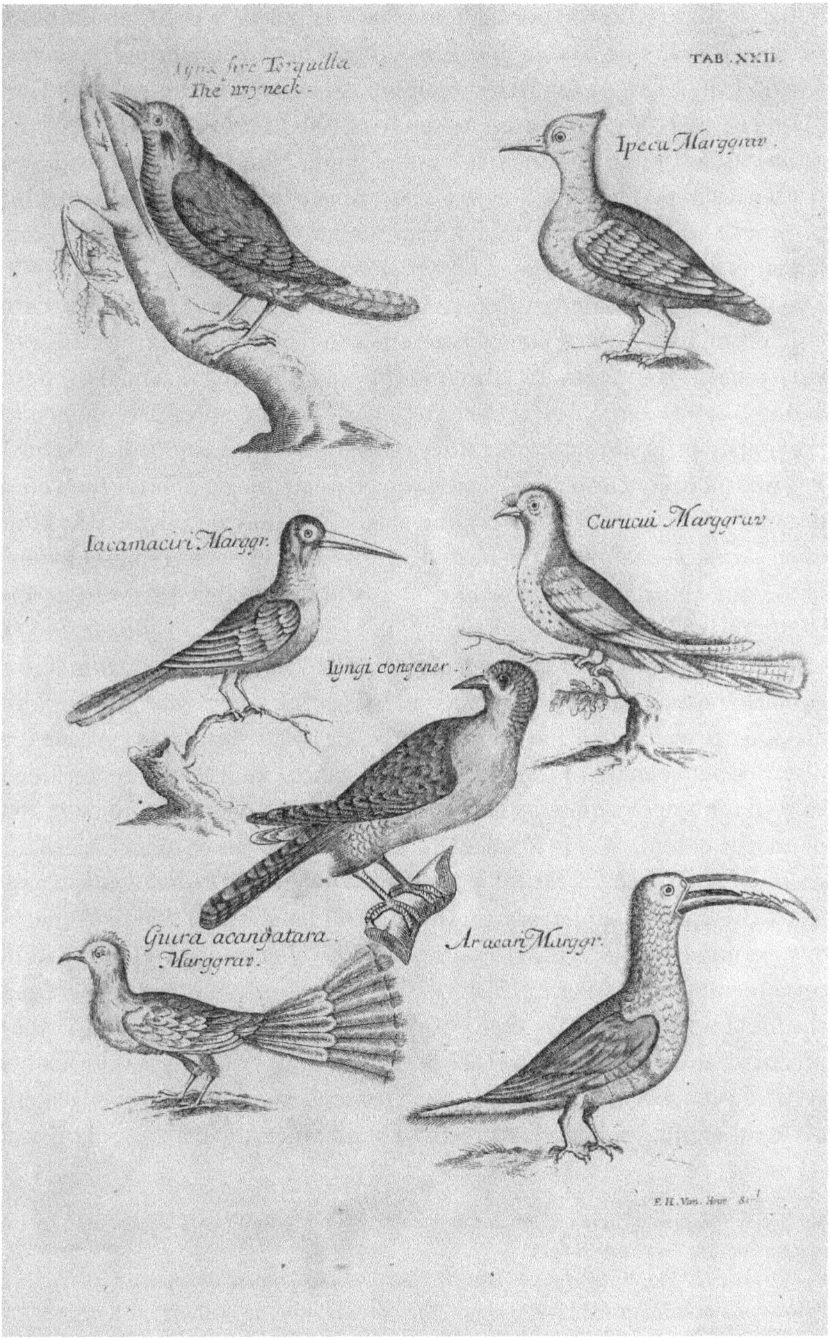

Figure 3.1: Frederick van Hove, *Plate 22*, Engraving, from Francis Willughby and John Ray, *The Ornithology of Francis Willughby* (London: Printed by A.C. for John Martyn, 1678). Cullman Library, Smithsonian Libraries via the Biodiversity Heritage Library.

indiscriminately in the period.[7] Sachiko Kusukawa, writing about Ray's *Historia Plantarum*, states "Ray himself was not so concerned, it seems, whether they were taken from life or not, since even for his own work on plants he rejected the phrase 'taken from the life,' because most of the descriptions, he said, were taken out of books."[8] In the letter Kusukawa references in this passage, Ray was chastising a friend for encouraging his printer to add the phrase "taken from the life" to one of his descriptions of plants because it was based on a previously printed description. While Kusukawa is interested in the relationship of Ray's text to the life, I am expanding her work by considering the concept of "taken from the life" with regards to images. By studying the visual source material used for Willughby's *Ornithology*, it becomes clear that Ray gave precedence to previously printed materials rather than drawings made "from the life." Ray was more concerned with using agreed upon images of birds than ones drawn from specimens. The regime of accuracy Ray mobilized was built upon extensive reading and repeated comparisons between representations of the same bird. The enactment of judgment Ray modeled grew from this comparative procedure.

Examining the visual source material used by Ray provides an in-road for exploring what it meant in the second half of the seventeenth century for an image to be life-like. Ray approximated life by attempting to collapse the distance between his objects of study and printed images of them included in Willughby's *Ornithology* through his careful collection and collation of images of birds. Using Ray's attitude toward images as expressed in his writings along with the resulting plates in Willughby's *Ornithology* as a case study, this chapter explores larger issues about the status of "life-like" images and examines the differing attitudes toward printed images and drawings in seventeenth-century natural history adding to the extant literature on "from the life" by reorienting the conversation to stress *reading* as an important precursor to *observing*. In the case of Willughby's *Ornithology*, the effect of accuracy was established through visual references within the printed materials to other images. Furthermore, this chapter investigates the authority of

7 See, for instance: Parshall, "*Imago Contrafacta*," 554–579; and Swan, "*Ad vivum*," 353–372 and Kusukawa, "*Ad vivum* Images."

8 Kusukawa, "The *Historia Piscium* (1686)," 186. She bases this statement on a passage in a letter from Ray to Robinson (22 May 1685): "Mr. [Henry] Faithorne writes me, that you advised to add to the Brief Account I sent him of the H.P. [*Historia Plantarum*], these words, *as also a particular description of their Parts, as Roots, leaf, flower, seed etc. taken from the life;* which I cannot allow, it being not true, the far greater number of descriptions being transcribed out of books..." *Further Correspondence of John Ray*, 147.

different sets of images to show that the authority of the images in a single text is built upon other printed books. Willughby's *Ornithology* presents an aggregated form of authority gleaned from other texts and witnesses.

Ray's reliance on printed materials did not preclude the usefulness of first-hand examination of his objects of study. On the contrary, in an attempt to make his book the most useful to his readers, Ray took it upon himself not only to read the writings of earlier authors, but also to examine as many of the birds as possible himself. To make clear to his readers which birds he had seen himself, those birds he had not observed in person were marked with an asterisk. He notes that Ulisse Aldrovandi (1522–1605) and Konrad Gessner (1516–1565), the two great naturalists of the sixteenth century, "are for the most very accurate and full" when they are describing specimens that they saw themselves (Preface, n.p.). However, descriptions based on others' observations "are of a different stamp." Ray then set out to correct the errors made by earlier writers who had relied too heavily on second-hand testimony, which had led to a multiplicity of species because of vague descriptions. Because Aldrovandi and Gessner, as well as many other authors, included descriptions of every bird that was sent to them and reported by their predecessors going back to Aristotle, Ray found many duplicates among their accounts. In a letter to Martin Lister, Ray described his methodology as well as his reservations about the quality of his source material: "I have carefully studied Gesner and Aldrovandi, and have attentively compared their descriptions with ours. I am uncertain about a very small number, and they were clearly either not written by them at all or with insufficient care, so that they can be clearly distinguished from those which are the same as ours."[9] Inadequate descriptions led Ray to perform a selective process that aimed to eliminate these inherited ambiguities, and in that sorting, his own observations were key to the process. Reading and observing were parallel paths that had to be traveled together in order to reform natural history.

His efforts to create a collection of useful images that would further the study of birds did not go unnoticed. The Royal Society, which granted its imprimatur for the publication of the Latin edition on June 24, 1675, expressed its appreciation of the work Ray had done in completing Willughby's project and praised "the Industry, Care and Accuracy of the said person [Ray] in digesting and perfecting it [Willughby's text]," in a review of the book in the Society's *Philosophical Transactions*.[10] Ray's judgment and

9 John Ray to Martin Lister, 22 September 1667, as quoted in Birkhead et al., "Willgubhy's Ornithology," 279.

10 Oldenburg, "Account of Some Books," 481.

careful actions were recognized and praised. The act of digesting, that is, choosing what is useful and what is waste, yielded accuracy.

The *Ornithology* was to be but the first instalment in a larger project intended to systematize the entire animal kingdom. After Francis Willughby's death in 1672, Ray was named as one of his five executors, provided with an annuity of £60, and soon began putting together Willughby's research for publication.[11] In addition to praising Ray's work, the reviewer also praised Emma Willughby, Francis's widow, for her efforts and encouraged her to continue to support Ray's work by financing the plates that would hopefully accompany further volumes.[12] Despite the expense, intaglio printed images were seen as essential to this larger project.

The Ornithology of Francis Willughby began as a collaboration between Willughby and his tutor, John Ray, that aimed at reforming all of natural history with Willughby focusing on the animals and Ray on the plants.[13] The work was incomplete at the time of Willughby's death in 1672, and out of a sense of duty Ray credited Willughby on the title page of both the Latin and English editions.[14] In his biography of Ray, Charles Raven argued that Ray did the bulk of the work as Willughby had left only notes.[15] Raven's view has been challenged by Tim Birkhead, who organized a network of researchers to re-evaluate Willughby's contributions to the project.[16] Work was already underway on the text and plates in 1673 when Ray, in a letter to Martin Lister, wrote:

> I am going on as fast as I can with the *Ornithology*. That the Work may not be defective, I intend to take in all the Kinds I find in Books which Mr. *Willughby* described not, and to have a Figure for all the Descriptions

11 Mandelbrote, "Ray, John."

12 Unfortunately for the Royal Society, this was not to be the case. Instead of Emma Willughby financing the cost of the illustrations for Willughby's *Historia Piscium* (1686), the Royal Society took up a subscription among its members to fund the engraving of the plates. For a discussion of this book see: Kusukawa, "*Historia Piscium*," 179–197.

13 Derham, *Select Remains*, 48.

14 The full title of the Latin edition is: *Francisci Willughbeii, De Middleton in agro Warwicensi, Armigeri, E Regia Societate, Ornithologiæ Libri Tres*. Ray is credited further down on the title page: "Totum opus recognovit, digessit, supplevit Johannes Raius." The full title of the English edition is: *The Ornithology of Francis Willughby of Middleton in the County of Warwick, Esq.; Fellow of the Royal Society. In Three Books*. Ray's credit line is simply: "By John Ray, Fellow of the Royal Society."

15 Raven, *John Ray*, 308. Raven also outlines the arguments made by both sides in the debates about the amount of credit due to each of them. Raven, *John Ray*, 334–336.

16 Birkhead, ed., *Virtuoso by Nature*.

> I can procure them for. I have sent this Week to Mr. *Martin* to begin to get some Figures engraved.[17]

The "Mr. Martin" to whom Ray refers is John Martyn, who was Printer to the Royal Society and had also printed Hooke's *Micrographia*. Martyn published both the Latin edition of the book in 1675/6 and the English edition in 1678.[18] This passage indicates that there was simultaneous progress on both the text and the plates.

To produce the quantity and quality of plates that Ray required, Martyn hired at least three different engravers. Six of the eighty plates are signed by three different engravers: William Faithorne (ca. 1620–1691), William Sherwin (ca. 1645–after 1709), and Frederick van Hove (ca. 1628–1698).[19] Seventy-eight of the plates combine views of as many as seven birds on a single plate. The final two plates illustrate the discussion of methods for trapping different varieties of birds, which was added to the English edition. The text included in Willughby's *Ornithology* provided not only a view of the natural history of birds that had been cleared of the clutter of repeated species and fabulous birds, but also a new system for organizing birds that relied on form rather than function as had been the norm since antiquity. The plates for their part brought together views of birds that Ray believed to be the most like the life.

In examining what it meant to be most like the life, this chapter explores the choices Ray made as he digested the available visual and verbal materials. This exploration reveals that there were different pathways by which an image could accrue authority. Ray regarded images created by the printing press, the pen, and through the aid of the collected specimens in cabinet and aviary as all having value for his goal of presenting "the best and truest" images.[20] Ray studied representations of birds produced through all three of these mechanisms and in many cases, those created by the press were prioritized. Examining cases where Ray had multiple images of the same

17 John Ray to Martin Lister, 29 November 1673 in *Correspondence of John Ray*, 105.

18 Derham states that the book was published "in the later End of the Year 1675." (Derham, *Select Remains*, 46) The imprimatur is dated 24 June 1675 and the title page is dated 1676. Ray was already at work on an English edition by February 1675 when Martin Lister wrote to him with suggestions for additions to the new edition. Lister to Ray, 8 February 1675 in *Correspondence of John Ray*, 116–118.

19 For Faithorne, see plates 39 and 44. For Sherwin, see plate 48. For van Hove, see plates 12, 22, and 45.

20 My thoughts on the importance of these three categories are informed by: Johns, *Nature of the Book*; Freedberg, *Eye of the Lynx*; and Findlen, *Possessing Nature*.

bird clarifies our understanding of this process and the balance he sought between reading and observing to produce the effect of accuracy for his readers.

Ray's method for correcting the natural history of birds relied on printed books, preserved specimens, live specimens, and drawings by others. Ray saw his task as being twofold: first to sort through the existing literature to expunge the unnecessary duplication of species, and second to present this culled group of birds in a more organized and systematic fashion than those before him. In this task, he relied on the form of the bird as the determining factor. Ray's project was not a wholesale rejection of the ancients, but rather an attempt to winnow out the errors and duplications that had entered into the corpus of knowledge regarding the natural history of birds. In the images as well as the text Ray sought to reproduce only the best: "Of the same Species of Bird when more Figures than one occurred either in divers Authors, or our own Papers, or both, we caused only one, which we judged to be the best to be engraven" (Preface, n.p.). The image Ray found "to be the best" was not necessarily the closest to a live specimen, but rather the one that was most recognizable to his readers. Much like Hooke sought to cultivate "a *faithful Eye*," Ray exercised his own visual judgment to distinguish which images were the best.[21] Following the methods propounded by Francis Bacon, Ray, then, had examined and edited both the printed and manuscript sources available to him to produce a book that did not contain any "dubious *Icons*" or "*Hieroglyphics, Emblems, Morals, Fables…*" only "what properly relates to their [birds'] Natural History" (Preface, n.p.). Ray's careful handling of information and decisions about what was properly part of a natural history of birds worked to make his mediation seem like less of an intervention. The care he took in selecting the images helped to establish the appearance of accuracy within his book, which was recognized by Henry Oldenburg in his review of the book discussed earlier.

The images included in the book were crucial to Ray's larger goals for the project. He stated in the Preface that he wanted the book to be useful "to the assistance and ease of those who addict themselves to this most pleasant, and no less useful part of Philosophy" (Preface, n.p.). To that end, he saw the figures as being essential to helping others identify species they encountered while in the field and traveling. His method for identifying a bird began with referring to the figures: "Nor will it be difficult to find out any unknown Bird that shall be offered: For comparing it with the Tables

21 Hooke, *Micrographia*, "The Preface," a2v.

first...the Bird may soon be found" (Preface, n.p.). The "Tables" or plates of illustrations of birds were then critical to his method for furthering the study of the natural history of birds.

Examining the visual materials available to Ray for four specific birds highlights the series of decisions that he made in determining the view of a bird that was "most like the live Birds" and uncovers the multivalent status of authority in early modern natural history. On the title page and in the Preface, Ray refers to the figures first as "most Elegant Figures, nearly resembling the live BIRDS, Engraven in LXXVII Copper Plates" (title page) and then as "the best and truest, that is, most like the live Birds, of any hitherto engraven in Brass" (Preface, n.p.). He claims expressly that the plates in his book are the best that are available. By examining the cases of the dodo, the golden eagle, the smew, and the grey gull, I want to highlight the complexity of the choices Ray made and show that there was a not simple answer to which image is "the best and the truest."

Resembling the Text: The Dodo (Raphus cucullatus)

Ray's process for describing and illustrating the dodo highlights the complexity of studying exotic species, as a great deal of attention was paid in the period to a bird about which there was a finite amount of information available. In the 1670s, as John Ray was assembling the text for Willughby's *Ornithology*, the dodo, solely found on the island of Mauritius, had been known to natural historians for approximately seventy years, and further, had not yet been confirmed as being extinct.[22] There was not a lot known about them and naturalists who had been to Mauritius wrote none of the accounts in circulation. Instead, naturalists depended on the accounts of sailors and merchants who had traveled to the island.

While it was not uncommon in this period for a bird to be known by multiple names, in the case of the dodo the names circulating were quite different and reflect a lack of consensus about the nature of the bird and how to classify it. John Ray cited three sources for the various names in the existing literature that refer to the bird that was commonly called the dodo (now *Raphus cucullatus*): "The Dodo, called by Clusius Gallus gallinaceus peregrinus, by Nieremberg Cygnus cucullatus, by Bontius

22 For thorough accounts of the documentation of the dodo see: Hume, "History of the Dodo," 65–89 and Parish, *Dodo and the Solitaire.*

Dronte" (153). His sources then were *Exoticorum libri decem* (Leiden, 1605) by Carolus Clusius (1526–1609), *Historia naturae* (Antwerp, 1635) by Juan Eusebio Nieremberg (1595–1658), and *De medicina Indorum* (Amsterdam, 1658) written by Jakob de Bondt (1592–1631) and contained within Willem Piso's *De Indiae utriusque re naturali et medica*.[23] Clusius thought that the bird should be allied with chickens (*Gallus*), whereas Nieremberg argued with his classification that it should be considered with the swans (*Cygnus*). Bondt skirted the debate by only providing a Dutch vernacular name.[24] Ray categorizes the dodo with neither of these groups, but rather with other flightless birds, in keeping with his categorization of birds based on form.[25]

In his evaluation of the early sources of information about the dodo, Julian Hume argues that at most three or four whole birds were successfully brought back to Europe.[26] Ray had seen a complete specimen: "We have seen this Bird dried, or its skin stuft in Tradescants Cabinet" (154). John Tradescant the elder (d. 1638) established the collection that would become the Tradescant's Cabinet at his house in Lambeth in 1628 after a decade of collecting.[27] The Tradescants allowed visits not only from dignitaries and scholars but also from the public, making it rare among cabinets of curiosities in the seventeenth century.[28] At the Tradescant's, Ray saw prepared specimens of a dodo, the Indian Mockbird, "a red Indian bird," the red-breasted Indian Blackbird, a penguin, and a puffin.[29] It is thought

23 William Poole has traced the copy of Nieremberg Ray used to a 1994 sale catalogue. Poole, "The Willughby Library," footnote 25, 235.

24 For a discussion of Bondt's time in the East Indies see: Cook, *Matters of Exchange*, chap. 5, "Truths and Untruths from the Indies," esp. 190–200.

25 The group includes the ostrich, emu, and dodo and is called: "The greatest Land-birds, of a peculiar kind by themselves, which by reason of the bulk of their bodies, and smalness of their Wings cannot fly, but only walk," 149.

26 Hume, "History of the Dodo," 66. Joylon Parish also arrives at a total of four whole specimens having arrived in Europe. Parish, *Dodo and the Solitaire*, 175.

27 MacGregor, "Cabinet of Curiosities," 149.

28 MacGregor, "Cabinet of Curiosities," 150.

29 In reference to the Indian Mockbird, "We saw this Bird dried in Tradescants Cabinet," 193. In the account of the "Aldrovandus his Brasilian Merula": "We have seen in Tradescants Cabinet a red Indian bird dried, of the bigness almost of a Mavis, having a long Tail, which perchance is the same with the bird in this Article described," 194. In reference to the red-breasted Indian Blackbird, "We saw the Case of this bird in Tradescants Cabinet," 194. In the account of "The Bird called Penguin by our Seamen, which seems to be Hoiers Goifugel": "I saw it also in Tradescants Cabinet at Lambeth near London," 322. In reference to the puffin, "Those which I saw dried in the Repository of the Royal Society, and in Tradescants Cabinet, seemed to me somewhat bigger," 333.

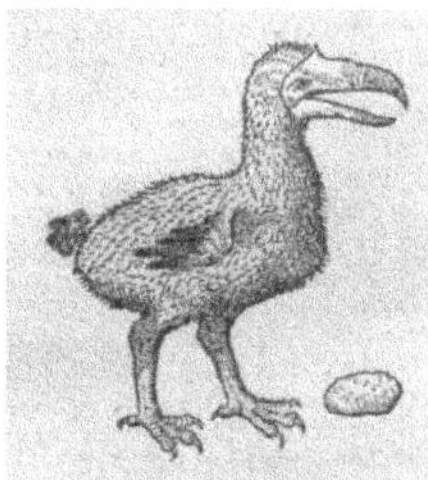

Figure 3.2: Anonymous, *Gallus gallinaceus peregrinus*, page 100, Woodcut, from Carolus Clusius, *Exoticorum Libri Decem* (Antwerp: Ex Officinâ Plantinianâ Raphelengii, 1605). Missouri Botanical Garden, Peter H. Raven Library via Biodiversity Heritage Library.

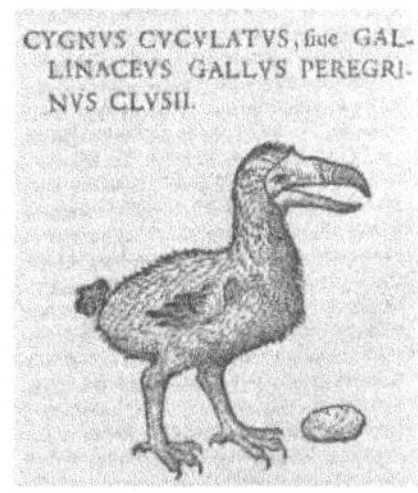

Figure 3.3: Anonymous, *Cygnus cuculatus*, page 231, Woodcut, from Juan Eusebio Nieremberg, *Historia Naturae, Maxime Peregrinae, Libris XVI* (Antwerp: Ex officina Plantiniana Balthasaris Moreti, 1635). Cullman Library, Smithsonian Libraries via the Biodiversity Heritage Library.

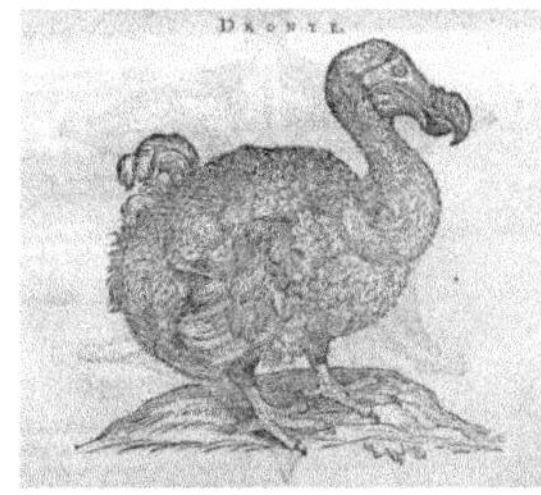

Figure 3.4: Anonymous, *Dronte*, page 70, Woodcut, from Jakob de Bondt, "Historiæ Naturalis and Medicæ Indiæ Orientalis Libri Sex," in *De Indiæ Utriusque Re Naturali et Medica Libri Quatuordecim*, edited by Willem Piso (Amsterdam: Apud Ludovicum et Danielem Elzevirios, 1658). Missouri Botanical Garden, Peter H. Raven Library via Biodiversity Heritage Library.

that Tradescant's dodo was alive in London as early as 1638, and it was included in Hume's count of three or four specimens making it safely to Europe before the bird's extinction in the 1690s.[30] While Ray had seen a dried specimen of a dodo, he had not seen a live one and therefore based his account in large part on these three previously printed sources.[31] Clusius had also not seen a live bird but based his account on the writings of Dutch mariners and on his viewing of the leg of a dodo. Ray included Clusius's entire description, which mentions the leg directly as well as the owner of the leg, Pieter Pauw (153). Clusius based his account on a second-hand description of a live bird and his own viewing of a part of a bird in the collection of a prominent colleague.[32]

All three of the accounts cited by Ray are accompanied by woodcut illustrations of the bird. (Fig. 3.2–4) The source for Clusius's depiction of the dodo was the journal of a Dutch admiral, Cornelis Jacob Van Neck, who wrote

30 Hume, "History of the Dodo," 69, 77.

31 Although the description of this bird is not marked with an asterisk to indicate that Ray had not seen a live specimen, it is included in a list of errata at the beginning of the book. "The Birds marked with an Asterisc are such as we our selves saw not, nor described, but borrowed their descriptions of others," Errata, n.p.

32 Hume believes this leg to have been "an early arrival from Mauritius." Hume, "History of the Dodo," 77.

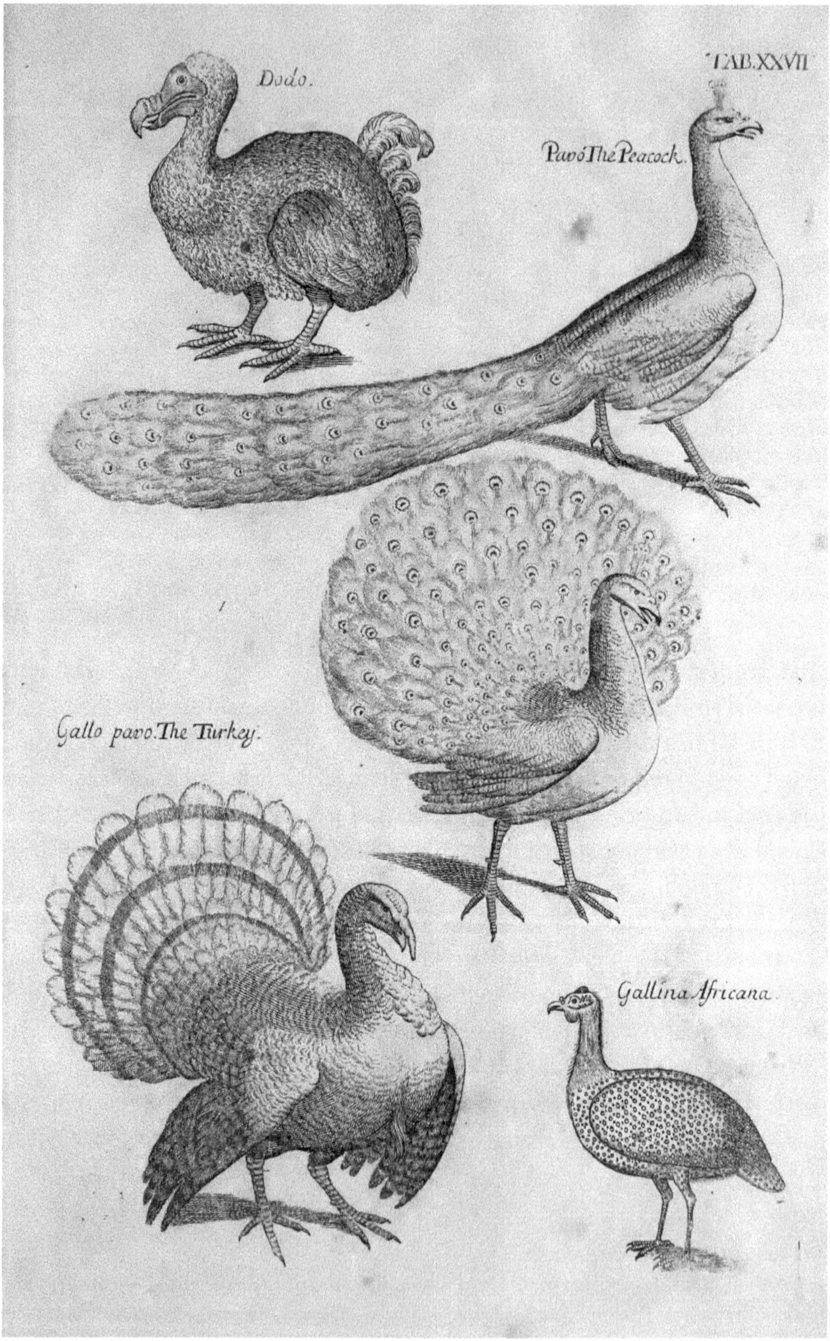

Figure 3.5: Anonymous, *Plate 27*, Engraving with etching, from Francis Willughby and John Ray, *The Ornithology of Francis Willughby* (London: Printed by A.C. for John Martyn, 1678). Cullman Library, Smithsonian Libraries via the Biodiversity Heritage Library.

the first account of the dodo in 1598.[33] Nieremberg's illustration is in turn based on Clusius's, although he provides a different name for the bird along with Clusius's name. The illustrations in Clusius and Nieremberg only provide the bare indicators of the features described by Clusius, as quoted by Ray:

> They reported that it is covered with thin and short feathers, and wants Wings, instead whereof it hath only four or five long, black feathers; that the hinder part of the body is very fat and fleshy, wherein for the Tail were four or five small curled feathers, twirled up together, of an ash-colour (153).

Clusius and Nieremberg's images clearly show the four or five feathers that make up the wings and the small, curled feathers that compose the tail, but the images do not provide a sense of the body being "very fat and fleshy." Bondt's illustration on the other hand fits more closely with the descriptions provided by him and Clusius. It is this illustration that Ray used for Willughby's *Ornithology.* (Fig. 3.5) The illustration accompanying Bondt's account of the dodo appears to be based on a painting by Roelandt Savery (1576–1639).[34] (Fig. 3.6)

In selecting an image to accompany his description of the dodo, Ray had access to three printed descriptions of the bird, two printed images, and one stuffed skin. In this case "the best and the truest" image was copied from a description of the bird in an account of the East Indies that Willem Piso thought would be helpful to Dutch mariners and merchants. Ray selected from among the available images the one that most closely resembled the written description, as he had no live bird with which to compare it.

Resembling the Printed Record: The Golden Eagle (Aquila chrysaetos)

Examining Ray's description of the golden eagle (*Aquila chrysaetos*), along with the available images of this bird, reveals the value placed not only on printed

33 Hume, "History of the Dodo," 73 and Parish, *Dodo and the Solitaire*, 107–109. Parish translates the relevant passages from Clusius into English.

34 This conclusion is based both on the visual similarity of the two images and Hume's statement: "It is from this Dodo painting that most post 1638 Dodo illustrations are derived." Hume, "History of the Dodo," 74. This painting is currently in the collection of the Natural History Museum in London and was once owned by Sir Hans Sloane. Thackray, *A Catalogue of Portraits, Paintings, and Sculpture*, 51–52.

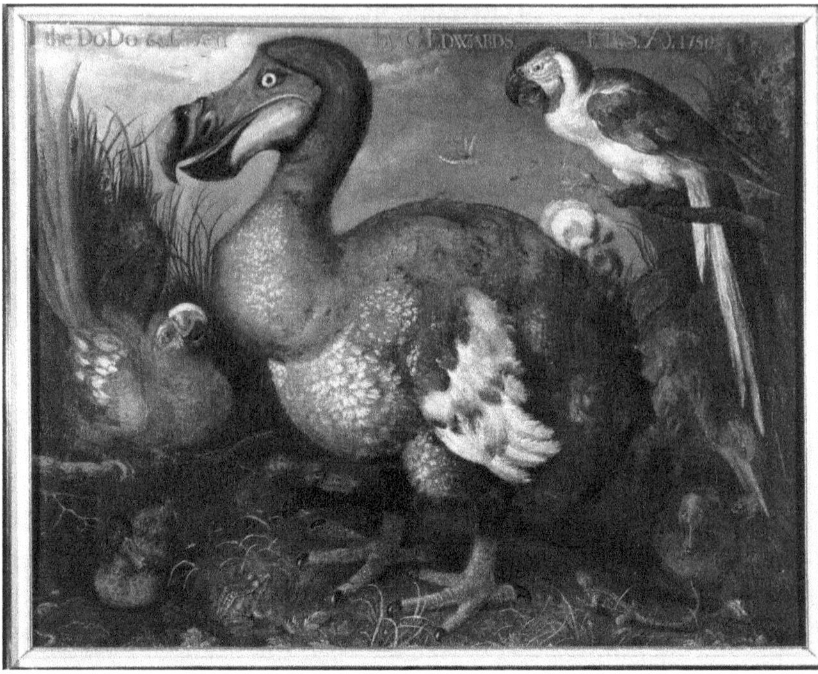

Figure 3.6: Roelandt Savery, *Raphus cucullatus (dodo), Ara macao (scarlet macaw)*, Oil on canvas, c. 1626. © Trustees of the Natural History Museum, London, #9391.

images, but also on Aldrovandi's in particular, highlighting that "the best and the truest" was often the one closest to a reliable source, not necessarily a live bird. Ray used a previously printed image even though he had access to the live bird and to a drawing of one. Further, this example shows that Aldrovandi's images of birds were valued more highly by Ray than those of Konrad Gessner, who published his own encyclopedia of animals, including birds, beginning in 1555.[35] Gessner's *Historia animalium* is an encyclopedic treatment of quadrupeds, birds, fish, reptiles, and insects, each volume of which is arranged alphabetically. Ray's reliance on Aldrovandi stems from his understanding of his working methods as well as the overall quality of his text and images.

Aldrovandi's book was by far the most heavily relied upon by Ray and the engravers. He based the text and images of his *Ornithologiae* on his own observations, accounts sent to him by correspondents, and printed texts as well as drawings and specimens that he acquired.[36] His collections were

35 Gessner, *Historiæ Animalium*.

36 Lind, *Aldrovandi on Chickens*, xxviii–xxix. For a discussion of Aldrovandi's extensive library see: Duroselle-Melish and Lines, "Library of Ulisse Aldrovandi," 133–161.

the basis for the museum he created in his house that he gave to the city of Bologna at his death.[37] Aldrovandi sought to create an encyclopedia of nature that he predicted would fill fourteen volumes; the *Ornithologiae* was the first part of this project to be published.[38] This massive book collected and presented all the knowledge that was circulating about birds at the end of the sixteenth century, and, much to later scholars' dismay, included a great deal of fabulous information. Ray made clear in the Preface that he sought to distinguish his work from that of Aldrovandi and his predecessor Konrad Gessner: "it was neither the Authors [i.e. Willughby's], nor is it my intention to write Pandects of Birds, which should comprise whatever had been before written of them by others, whether true, false or dubious, that having already been abundantly performed by Gesner and Aldrovandus" (Preface, n.p.). Willughby and Ray, in keeping with the Royal Society's Baconian project, sought to present a text based on their observations that sorted out the fabulous material that had accrued in the texts of Aldrovandi and Gessner. Ray then set out to improve upon Aldrovandi's text, but still relied heavily on his illustrations. Willughby and Ray brought the same careful attention to the visual material they examined as they did to the written record and their own observations.

Of the 373 images included in Willughby's *Ornithology*, 129 (i.e., 34.6 percent) were based upon images from Aldrovandi. Nearly 50 percent of the images based on printed sources are taken from Aldrovandi's *Ornithologiae*. In plate 29, for example, six of the seven birds are copied from Aldrovandi.[39] (Fig. 3.7) While this is a fairly large number to take from one source, it is important to note that this selection represents less than 20 percent of the total images included in Aldrovandi's massive three-volume work.[40] Ray was selective about which images he used from Aldrovandi and did not include many of the images that represented variants of species. Ray's efforts to cull out the duplications from Aldrovandi's work extended beyond the text to include the figures. That said, Aldrovandi's woodcut illustrations were a great source of information for Ray and the engravers. Ray knew that the images in Aldrovandi's printed book were based in part on drawings that Aldrovandi had collected in preparation for this own work because he and Willughby had visited the museum in Bologna while traveling in

37 Castellani, "Aldrovandi, Ulisse," vol. 1, 109.

38 Stresemann, *Ornithology*, 21.

39 The seventh is taken from Jacob Bondt's account of the East Indies published in Piso, *De Indiae*, 65.

40 In total, there are 687 illustrations accompanying Aldrovandi's text.

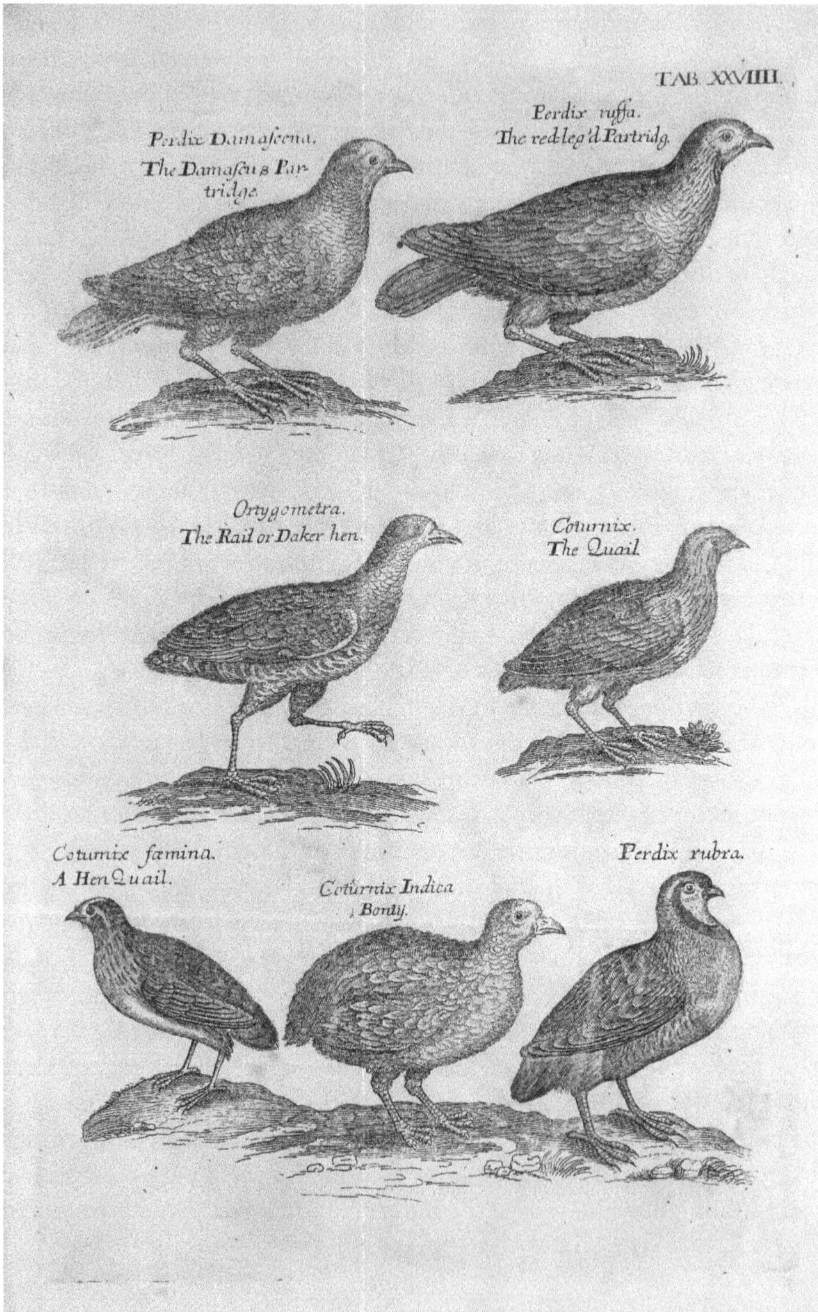

Figure 3.7: Anonymous, *Plate 29*, Etching with engraving, from Francis Willughby and John Ray, *The Ornithology of Francis Willughby* (London: Printed by A.C. for John Martyn, 1678). Cullman Library, Smithsonian Libraries via the Biodiversity Heritage Library.

Italy in 1664.[41] Ray's process for preparing Willughby's *Ornithology* was in fact similar to Aldrovandi's in that he was working from both printed and manuscript sources as well as specimens. Both Aldrovandi and Ray believed that illustrations were a necessary component of their projects.[42]

The bulk of Ray's description of the golden eagle was based on Alrovandi's account, and Ray noted that Aldrovandi had been sent one of these birds: "So far Aldrovandus: to whom this Bird was sent by the Great Duke of Tuscany" (59). At the end of his description of the golden eagle as reported by Aldrovandi, Ray appended a description of "The Golden Eagle with a white ring about its tail." He had seen four birds of similar description in London: "We saw three Birds of this sort in the Royal Theriotrophium near the Tower of London, and a fourth in St. James Park near Westminster" (59). After providing a detailed account of their characteristic traits, he concluded, "I take it to be specifically the same with the precedent [i.e. the Golden Eagle of Aldrovandi]" (59). Ray, then, had seen multiple living specimens of this bird but still based the bulk of his account on Aldrovandi. In this case, Ray's printed image approximated a previously printed source.

Although Gessner is often referenced in the text of Willughby's *Ornithology* and he is frequently listed as the authority for the name of a bird on the plates of illustrations, none of the images of birds that were included in his work were used as sources for Ray's plates. Instead, preference was given to Aldrovandi's illustrations. Comparing the images of the golden eagle from all three books makes it clear that Aldrovandi's image was based on Gessner's, and Ray's, in turn, was based on Aldrovandi's. (Fig. 3.8–10) These acts of recycling complicate our understanding of accuracy because Ray used an image without a clear relationship to the life by instead prioritizing an image from an authoritative printed source. While the image of the eagle is nearly identical in all three images, in Aldrovani's view of the bird a small piece of land with a plant with very round flowers or seedpods has been added. This plant is also present in Ray's plate. Even though Gessner's text was available to Ray and the image in Aldrovandi was clearly derivative, the image of the golden eagle in Willughby's *Ornithology* uses Aldrovandi as its source. Ray had access to Gessner's book,

41 "Heer, by the favour of Dr. *Ovidio Montalbano*, one of the Professors, we had a sight of the *Museum* of *Aldrovandus*, which by his last will he left as a Legacy to the City. It is kept in the Cardinal Legates Palace, commonly called *Palazzo del Confaloniero*. Among many natural and artificial Rarities therein preserved, we took more especial notice of 10 Volumes of the pictures of Plants, and 6 of Birds, Beasts, and Fishes, drawn exactly in colours by the hand." Ray, *Observations Topographical*, 234. Greengrass et al., "Science on the Move," 176–177.

42 Paula Findlen states that Aldrovandi "chastised the ancients for neglecting to illustrate their treatises." Findlen, *Possessing Nature*, 69.

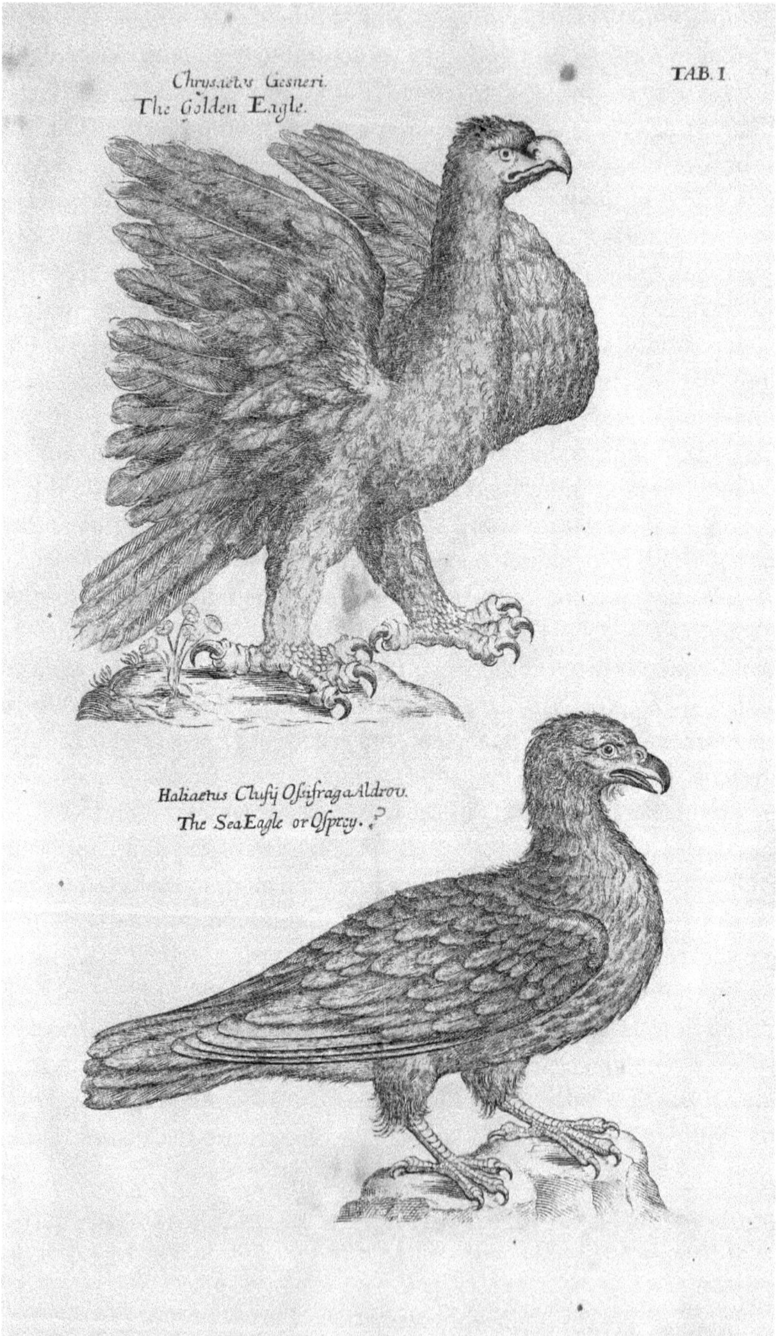

Figure 3.8: Anonymous, *Plate 1*, Etching with engraving, from Francis Willughby and John Ray, *The Ornithology of Francis Willughby* (London: Printed by A.C. for John Martyn, 1678). Cullman Library, Smithsonian Libraries via the Biodiversity Heritage Library.

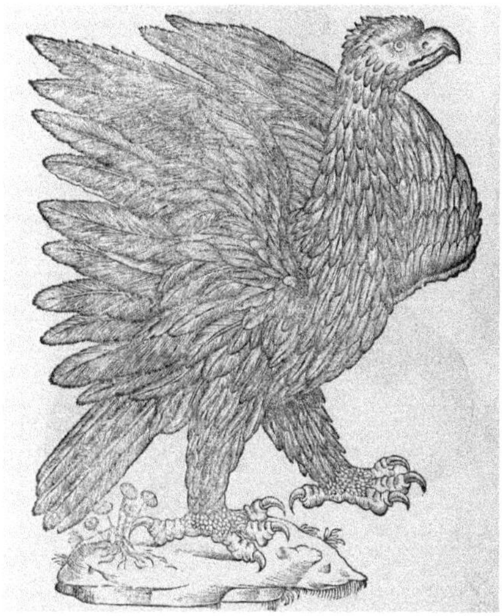

Figure 3.9: Anonymous, *Chrysaetos*, vol. 1, page 114, Woodcut, from Ulisse Aldrovandi, *Ornithologiae, Hoc est De Avibus Historiae Libri XII* (Bologna: Apud Franciscum de Franciscis Senensem, 1599–1603). Cullman Library, Smithsonian Libraries via the Biodiversity Heritage Library.

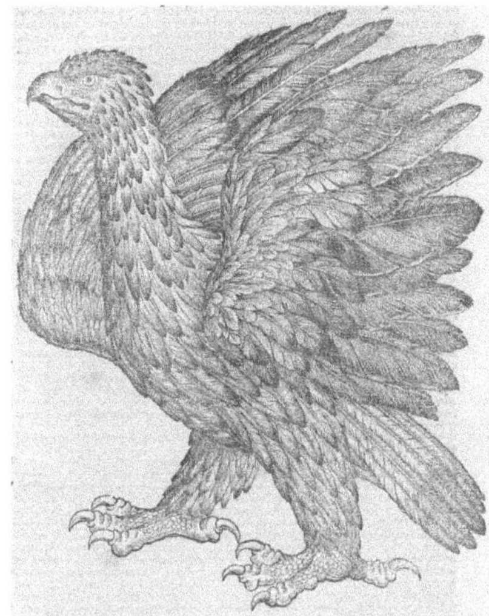

Figure 3.10: Anonymous, *Aquila Germana*, page 163, Woodcut, from Konrad Gessner, Historiæ Animalium Liber III (Zurich: C. Froschouerum, 1555). Getty Research Institute via the Biodiversity Heritage Library.

and since the images were printed with woodblocks and set with the type, the text could not have circulated without the images. Gessner is not, however, among the list of books that Ray cites as his sources for images for Willughby's *Ornithology* (Preface, n.p.). Aldrovandi's images are preferred over Gessner's because Ray placed higher value on Aldrovandi as a source for text and images. Further, this is in part due to Ray's awareness of Aldrovandi's collection of specimens and drawings.[43] Although Ray knew there were duplications and errors in Aldrovandi's *Ornithologiæ*, he continued to value his illustrations.

Furthermore, Ray valued Aldrovandi's illustration that had been copied from Gessner more highly than a drawing he had of a bird matching his description of "The Golden Eagle with a white ring about its tail." (Fig. 3.11) This drawing is part of a collection that consists of illustrations of 450 specimens of birds, fishes, and other water animals contained in 343 drawings.[44] The drawing is labeled simply "Eagle," but as Ray points out "Our

43 See Ray, *Observations Topographical*, 234.

44 My account of the collection is based both on my examination of it and on the description contained in Brown's online library catalog. For more extensive discussions of this collection see: Lownes, "Collection of Seventeenth-Century Drawings," 532–535; Lownes, *Renaissance Books*, 18–19; and Phillips, "Leonard Baldner," 332–341.

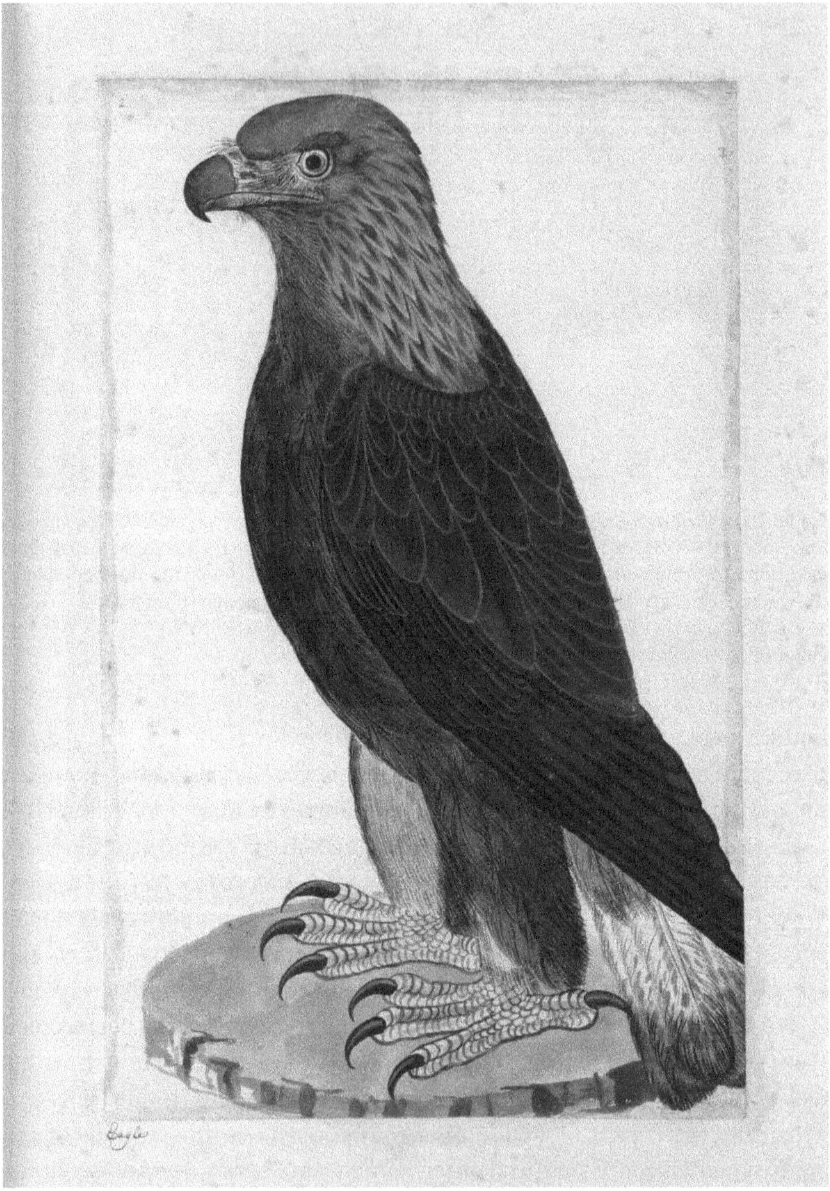

Figure 3.11: Anonymous, *Eagle*, vol. 2, fol. 2, Watercolor on paper, from *Recht natÿrliche Beschreibung und Abmalung der Wasser Vögel, Fischen, Vierfüssigen Thieren, Insecten, und Gewürm*, 1653–ca. 1663. Brown University, Hay Lownes Library, RARE 3-SIZE QL41 .B3 1653.

Country-men call this bird simply and absolutely the Eagle, without any Epithet of distinction" (59). This drawing clearly shows that "the Feathers covering the head and neck [are] not smooth and even, but rigid, narrow, and lying at a distance one from another" (59). This detail is quite exaggerated in the woodcut in Aldrovandi as it appears that the head and neck are covered with individual feathers that do not overlap. In addition, whereas the eagle in Aldrovandi's illustration appears to be strutting across the page, the one in the drawing sits on what seems to be a disk of wood and looks quite like a tamed bird of prey kept by the King of England. This drawing then appears to resemble a live bird but was not chosen as "the best and the truest" representation of this species.

As this example shows, Ray placed a greater value on images that were already circulating in print than those based on drawings. Although Ray had a drawing that matched his description of a live bird that was in a Royal Aviary, he chose to have the engraver use a previously printed woodcut of the bird as the source for the engraving in Willughby's *Ornithology*. I argue that this decision was motivated by Ray's desire to make the book useful. That is, it is more useful to include a recognizable view of a bird than one that is closer to the live bird that would have only been seen by the few who had access to the King's aviaries. In this case, "the best and the truest" image is not the one that most nearly resembles a live bird, but rather the one that is most easily recognized, allowing for clearer communication among a group of readers.

Resembling the Living: The Smew (Mergus albellus)

While the previous examples examined instances where prints were chosen as the sources for the illustrations in Willughby's *Ornithology*, the case of the smew or white nun highlights the use of manuscripts and direct observation in the production of the plates. The drawing that served as the source for the engraving of the smew in plate 64 is part of the album at the British Library that Willughby purchased from Leonhardt Baldner while in Strasbourg in July 1663.[45] (Fig. 3.12 & 13) Looking at this engraving and its source alongside the other information available to Ray about the bird shows how a drawing from a credible source was used to cull duplicate species out of the writings of Gessner and Aldrovandi.

In his outline of the visual material used in preparation of Willughby's *Ornithology*, Ray listed the volume purchased from Baldner and extoled his virtues as an apt source despite his low circumstances:

45 The drawings are Add. MS 6485. For details of Willughby's time in Strasbourg see: Greengrass et al., "Science on the Move," 195.

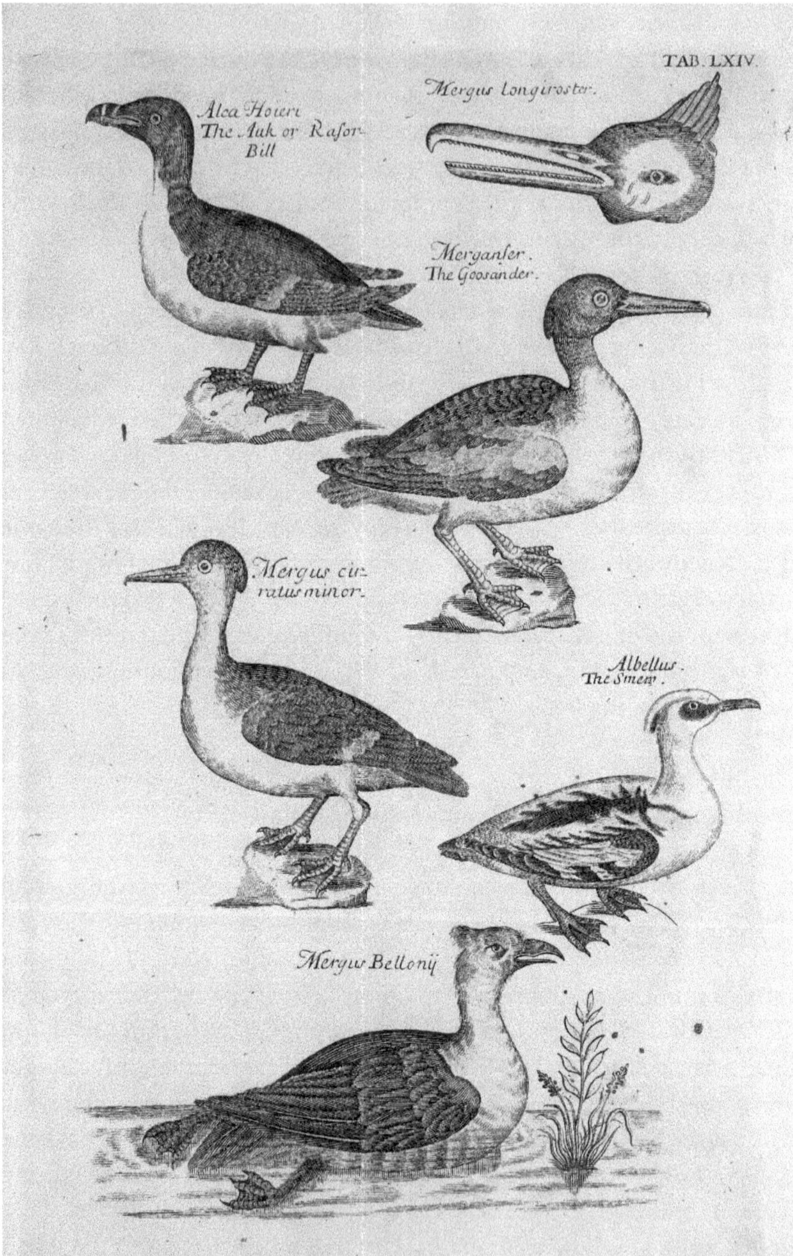

Figure 3.12: Anonymous, *Plate 64*, Engraving, from Francis Willughby and John Ray, *The Ornithology of Francis Willughby* (London: Printed by A.C. for John Martyn, 1678). Cullman Library, Smithsonian Libraries via the Biodiversity Heritage Library.

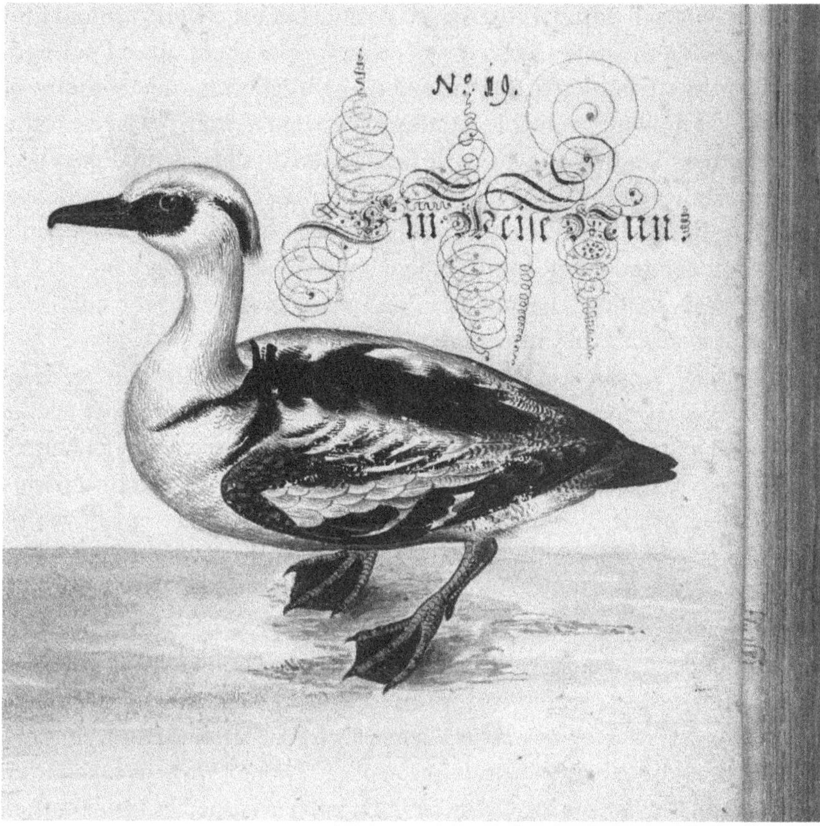

Figure 3.13: Leonhardt Baldner, *Ein Weisse Nunn*, Watercolor on paper, from *An exact natural Description and Deliniation of the Water-Fowls, Fishes, Fourfooted Beasts, Insects, and Worms* (Strassburg, ca. 1660). © British Library Board, Add. MS 6485.

> First, He [Willughby] purchased of one *Leonard Baltner*, a Fisherman of *Strasburgh*, a Volume containing the Pictures of all the Water-fowl frequenting the *Rhene* near that City, as also all the Fish and Water-Insects found there, drawn with great curiosity and exactness by an excellent hand. The which Fowl, Fishes, and Insects the said *Baltner* had himself taken, described, and at his own proper costs and charges caused to be drawn. Which curiosity is much to be admired and commended in a Person of his Condition and Education. For my part, I must needs acknowledge that I have received much light and information from the Work of this poor man, and have been thereby inabled to clear many difficulties, and rectifie some mistakes in *Gesner* (Preface, n.p.).

Baldner is praised both for his own curiosity about the world around him and for taking the pains to have the creatures he encountered painted. Ray's lengthy attribution of these drawings highlights his understanding of what makes drawings valuable for his project, that is, their "great curiosity and exactness." At the same time, he further clarifies his methodology and the importance of the "light and information" he received from Baldner. Rectifying Gessner's mistakes was not simply a textual problem, but a visual one. Judgment and connoisseurship go hand in hand in Ray's method. It was necessary to judge the quality of your sources as well as the quality of the visual materials they provided.

The case of the smew or white nun is an instance in which Ray "received much light and information" from one of Baldner's drawings. The mistake of Gessner's that Ray was able to rectify in this instance with the help of Baldner was in the duplication of species. Ray determined that the *Mergus Rheni* ("Diver of the *River Rhine*") of Gessner and the *Mergus major cirratus* of Gessner were the same as the Albellus of Aldrovandi and were one and the same as the smew (337–338). Ray was also able to correlate these descriptions with his observations of a smew that had been sent to him by a correspondent along with some fish and other waterfowl.[46] Ray's method relied on digesting printed text, printed images, drawings, and his own observation in order to present his readers with the "best and truest" images.

After providing Gessner's description of the *Mergus Rheni* as taken from Aldrovandi, Ray makes his claim for collapsing this species into another, citing Baldner as his authority:[47]

> I suspect that this Bird was no other than our *Albellus*, next to be described: Only the bigness and want of a crest forbid it. Perchance *Gesner* might describe it from the relation of others, or from a Picture. I am sure *Leonard Baltner*, a Fisherman and Fowler of *Strasburgh*, who did very diligently observe, gather together, and cause to be painted all the birds frequenting the *Rhene* thereabouts, gives us no other bird of this kind but the *Albellus*, to which also he gives the title of *White Nun* (337).

The conflict with Gessner's account and Ray's conception of the characteristic marks of the bird, the crest, is mediated by Ray's concession that Gessner

46 "With the fish I have put up in a box some waterfowl, viz., a Pocker, a Smew, three Sheldins, a Widgeon and a Whewer." Mr Dent to John Ray, 15 February 1674. *Correspondence of John Ray*, 16.
47 In the chapter heading for this description, Ray states that the name of the bird is from Gessner and then provides a page reference to Aldrovandi. "*Mergus Rheni* of Gessner, Aldrov. tom. 3. pag. 275," 337.

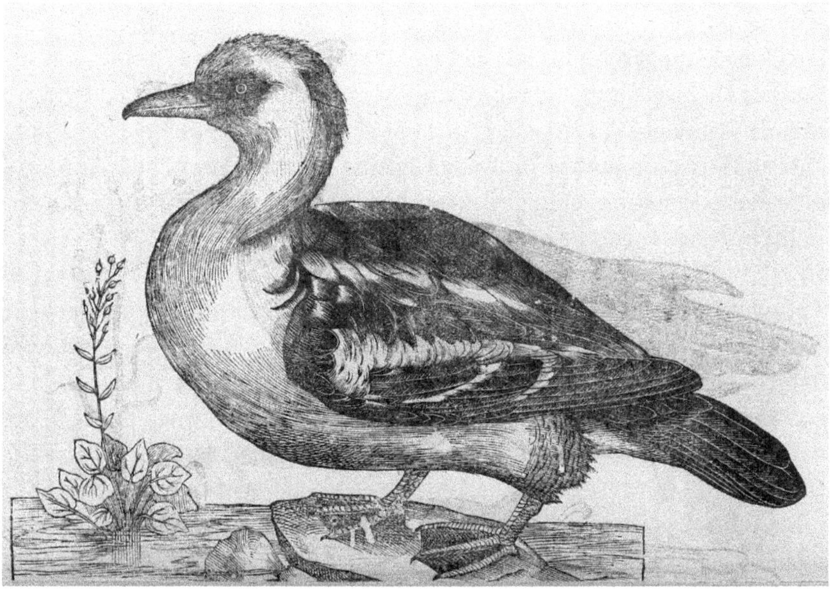

Figure 3.14: Anonymous, *Albellus aquaticus*, vol. 3, page 277, Woodcut, from Ulisse Aldrovandi, *Ornithologiae, Hoc est De Avibus Historiae Libri XII* (Bologna: Apud Franciscum de Franciscis Senensem, 1599–1603). Cullman Library, Smithsonian Libraries via the Biodiversity Heritage Library.

may have based his description only on his reading and not on any direct observation. The evidence given by Baldner is even more compelling then because it is based on his diligent observations. Baldner's credentials were again asserted when he was used as the source for an identification. In this passage, Ray reinforces our understanding of his comparative methodology.

Ray goes on to describe the *Albellus* based on his own observations and further argues that another species described by Aldrovandi should be collapsed into this bird:

> The *Albellus aquaticus* of *Aldrovand*, as it seems to me, differs not from this bird, for both the figure, and all the marks he gives of it, agree; only he makes no mention of the crest, perchance it was a young bird he described (338).

Here Ray cites not only Aldrovandi's account of the characteristic marks of the bird but also the figure he included. (Fig. 3.14) Instead of discounting Aldrovandi's account as being inaccurate because it was second-hand, as he had with Gessner's, he states that he might have been describing an immature specimen. The woodcut of Aldrovandi's *Albellus aquaticus* indeed bears a likeness to Baldner's drawing of the white nun, particularly in the coloring around the eye and the patterning of the wings. In his writing about the bird, Ray demonstrates his method for his readers as much as he presents his finding. His readers should take away an understanding of how to identify a smew and its place in a larger organizational scheme as well as an understanding of how to go about the process of sorting through the myriad sources of information a naturalist might encounter in his own identification process.

In this case, then, the evidence provided by "a Fisherman and Fowler of *Strasburgh*" is valued more highly than the two great natural historians of a previous generation. While in the previous example of the golden eagle, Aldrovandi was viewed as an authority, here his descriptions are discounted in favor of the combined effects of Ray's own observations of the bird and a drawing purchased from Baldner. Here the pen and brush of the artist have more authority than the press and the resulting image in Willughby's *Ornithology* approximates the likeness provided by a drawing instead of a print. Ray's process of culling duplicates and producing accuracy was aided by the collecting of drawings and books.

Resembling the Dead: The Great Grey Gull

A final example puts direct pressure on Ray's assertion that his figures resemble live birds because it in fact shows a bird that is clearly dead. Indeed, there are many instances in which the plates in Willughby's *Ornithology* show dead birds. In the case of the great grey gull on plate 66 the original drawing is extant.[48] (Fig. 3.15 & 16) Ray provided several variant names for the bird in the text: "The great grey Gull, which we take to be the Cornish Wagel, called at Venice, Martinazzo, at Amsterdam, the Burgomaster of Groenland: An Larus albo-cinereus torque cinereo of Aldrov?" (349).[49] The bird is labeled "*Larus griseus maximus*, The great grey Gull" on the plate and "Martinazo" on the drawing.[50] The inclusion of this bird among his "most Elegant Figures" raises important questions about the authority of images and the range of factors that influence the value of an image for natural history. As we saw in William Faithorne's depiction of a female cadaver discussed in the Introduction, this image has the visual traces of accuracy that would have been recognizable to, in this case, a natural historian who had studied dead specimens himself.

48 This drawing is part of the collection at Brown University, but it is not part of the group purchased from Baldner.

49 The question mark is included in the original.

50 There are other annotations on the drawings, but they appear to post-date the publication of Willughby's *Ornithology*.

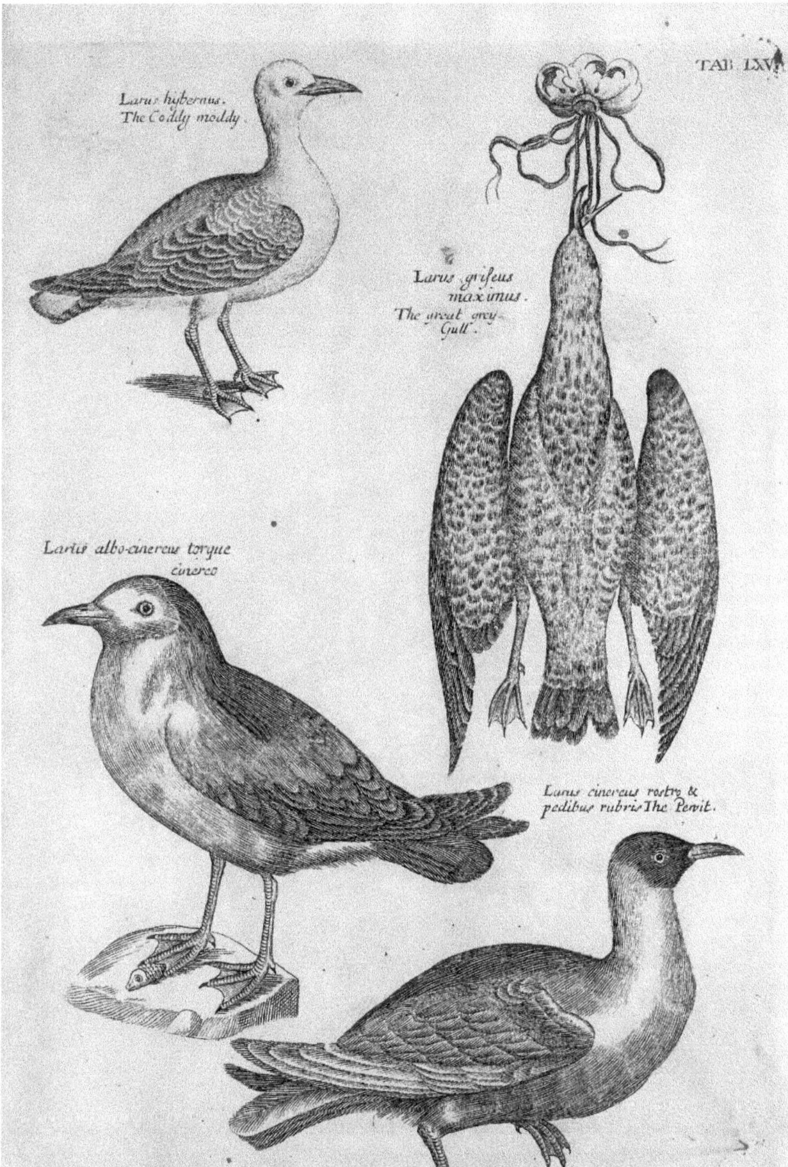

Figure 3.15: Anonymous, *Plate 66*, Engraving with etching, from Francis Willughby and John Ray, *The Ornithology of Francis Willughby* (London: Printed by A.C. for John Martyn, 1678). Cullman Library, Smithsonian Libraries via the Biodiversity Heritage Library.

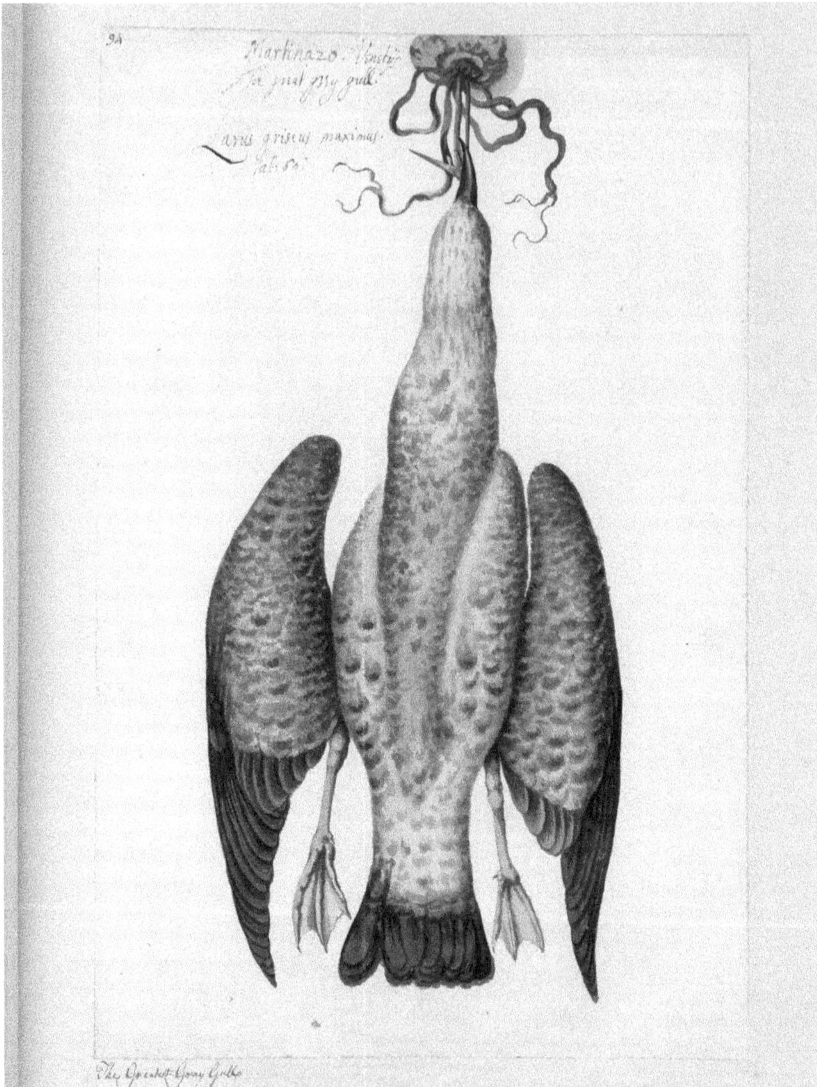

Figure 3.16: Anonymous, *The Greatest Grey Gull (Martinazo)*, vol. 1, fol. 94, Watercolor on paper, from *Recht natÿrliche Beschreibung und Abmalung der Wasser Vögel, Fischen, Vierfüssigen Thieren, Insecten, und Gewürm*, 1653–ca. 1663. Brown University, Hay Lownes Library, RARE 3-SIZE QL41 .B3 1653.

As with previous examples, the text related to this image presents evidence for the identification of the bird as well as evidence of Ray's working methods. The description of the "great grey Gull" makes clear that Ray described a specimen that he and Willughby saw and dissected.[51]

51 For a discussion of Ray's use of language to call to mind a vivid image, see: Alexander Wragge-Morley, *Aesthetic Science*, chap. 4, "Verbal Picturing," 106–134.

The text begins by giving a precise account of its size: "It weighed twenty two ounces; being stretcht out in length from the point of the Bill to the end of the Feet twenty one inches and an half, to the end of the Tail twenty one: its breadth was fifty three inches" (349). After providing a detailed report on the exterior characteristics of the bird, Ray proceeds to recount what they found upon opening the bird: "It hath a huge Liver, divided in two: a Gall annexed to the right Lobe. The Stomach more musculous than in carnivorous birds: The blind guts short and little, yet turgid, and full of Excrement" (350). In addition to lending credibility to his account of the bird, Ray's detailed description of the exterior and interior of the bird aligns his process with that of the anatomy theater. In describing the process as it occurred, he allows his readers to witness the dissection themselves.

First-hand observation and dissection were not always sufficient to definitively collapse multiple accounts of birds into a single identification. Ray concluded the account by positing that this bird "is very like to, if not the same with" the "*Larus albo-cinereus torque cinereo* of *Aldrovand*" (350). Plate 66 includes an image of the *Larus albo-cinereus torque cinereo* just below the great grey gull, but there is no separate description of the bird. The image in Willughby's *Ornithology* is taken from Aldrovandi's *Ornithologiae*. (Fig. 3.17) Given Ray's assertion in the Preface that he would not provide images of duplicate species and would only cause those "judged to be the best to be engraven," it appears that he was not convinced that these two images in fact represented the same bird (Preface, n.p.).

The great grey gull is an instance in which a drawing of a bird and a written description of a bird could not be reconciled. The drawing of the "Martinazo" and the engraving after it clearly show a dead specimen. This bird is hung from a wall with an elaborate bow and a ribbon that was tied through its beak. The contrast with the *Larus albo-cinereus torque cinereo* is dramatic. Where Aldrovandi's bird stands erect on a rock with a freshly caught fish under its foot, the great grey gull hangs limply with its wings and feet at improbable angles. Rather than make the bird conform to the expectations of the living as a taxidermy bird would, this bird conforms to the expectations of a dead bird. The splayed form and distinct ribbon simultaneously evoke Northern European images of dead game and Vesalius's cadavers held up by the hangman's rope. Ray was interested in both the living and the dead as sources of information for his book. This example highlights how there are times when an image of a dead bird is as close to the life as possible.

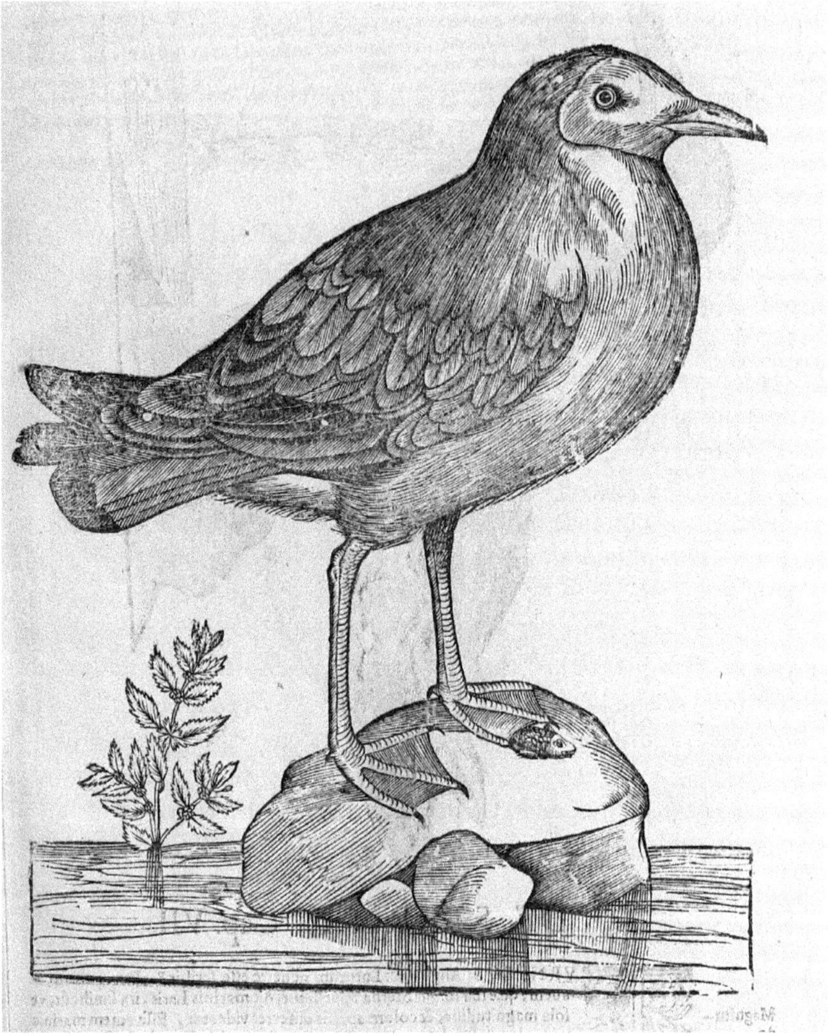

Figure 3.17: Anonymous, *Larus albocinereus torque cinereo*, vol. 3, page 77, Woodcut, from Ulisse Aldrovandi, *Ornithologiae, Hoc est De Avibus Historiae Libri XII* (Bologna: Apud Franciscum de Franciscis Senensem, 1599–1603). Cullman Library, Smithsonian Libraries via the Biodiversity Heritage Library.

Conclusion

These examples examine different scenarios in which Ray had multiple images of the same bird and had to choose which one best fit his purposes. Images were selected that were recognizable and presented the most authoritative view of the bird. Further, they were used to adjudicate differences among authors and to record particular specimens. Ray's project was a divided one because the

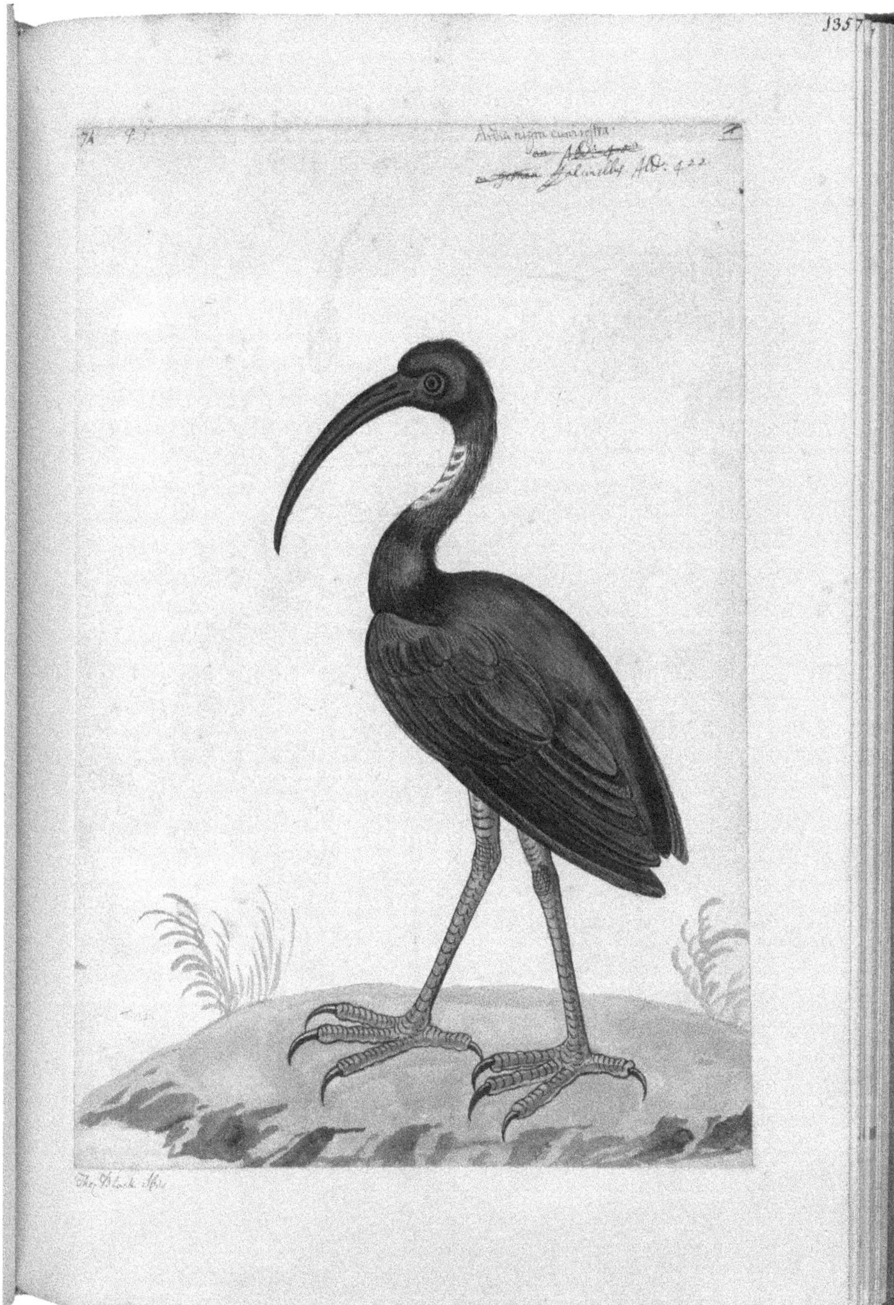

Figure 3.18: Anonymous, *The Black Ibis*, vol. 1, fol. 74, Watercolor on paper, from *Recht natÿrliche Beschreibung und Abmalung der Wasser Vögel, Fischen, Vierfüssigen Thieren, Insecten, und Gewürm*, 1653–ca. 1663. Brown University, Hay Lownes Library, RARE 3-SIZE QL41 .B3 1653.

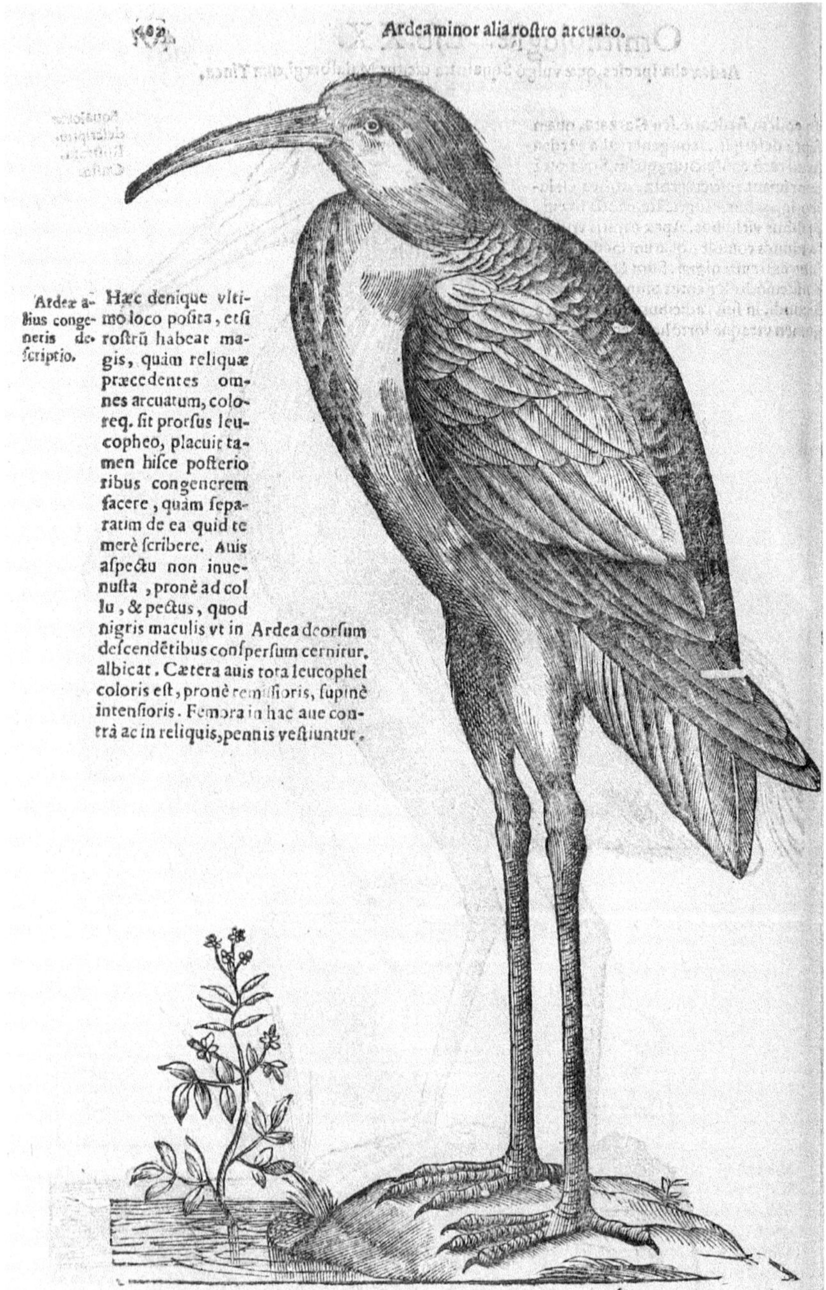

Figure 3.19: Anonymous, *Ardea minor alia rostro arcuato*, vol. 3, page 402, Woodcut, from Ulisse Aldrovandi, *Ornithologiae, Hoc est De Avibus Historiae Libri XII* (Bologna: Apud Franciscum de Franciscis Senensem, 1599–1603). Cullman Library, Smithsonian Libraries via the Biodiversity Heritage Library.

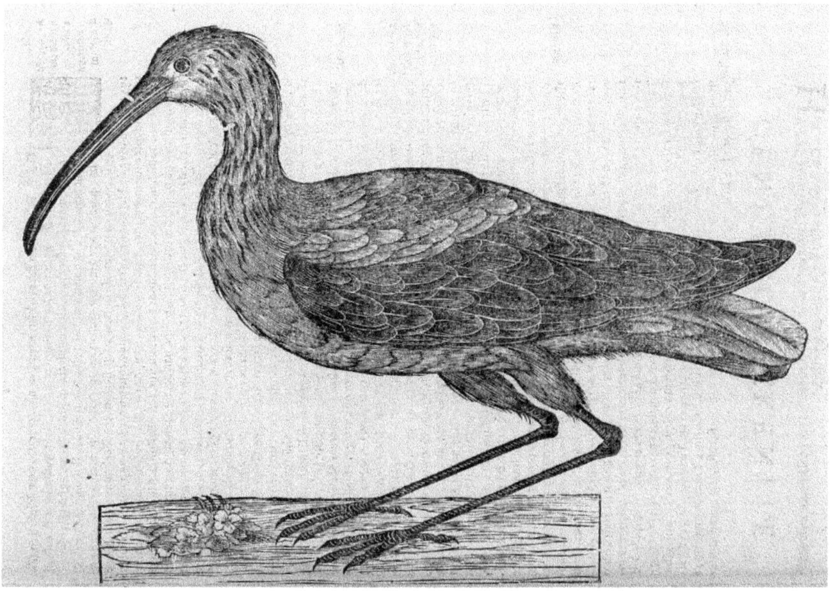

Figure 3.20: Anonymous, *Falcinellus Avis falcate*, vol. 3, page 423, Woodcut, from Ulisse Aldrovandi, *Ornithologiae, Hoc est De Avibus Historiae Libri XII* (Bologna: Apud Franciscum de Franciscis Senensem, 1599–1603). Cullman Library, Smithsonian Libraries via the Biodiversity Heritage Library.

path to achieving a proximity to life was unclear, as was the route to producing a collection of authoritative images. There was no single procedure to create accuracy. A drawing of a bird was not automatically accepted as "the best and the truest," rather drawings were cross-referenced with printed images as well as live and dead specimens. Ray carefully checked the images he and Willughby had collected against printed images and written descriptions to create a set of useful representations of the birds described in the text. In the Preface, Ray described their collecting practices and stressed the need for figures: "because elegant and accurate Figures do much illustrate and facilitate the understanding of Descriptions" (Preface, n.p.). His "accurate Figures" then were the result of his judgment of his materials and the care he put into collecting and collating the available visual materials. Reading and observing were paired processes. Both were necessary for the reformation of natural history.

Although Willughby actively collected and commissioned drawings of birds, he and Ray at times had difficulty correlating the drawings they had with previously printed accounts and images of similar birds. Looking at a drawing that was labeled in the mid-eighteenth century as a black ibis (probably a glossy ibis, *Plegadis falcinellus*) reveals the revisions Willughby made to his identification of the species represented in a drawing. (Fig. 3.18) An unidentified hand has labeled the drawing "*Ardea nigra curvirostra*"

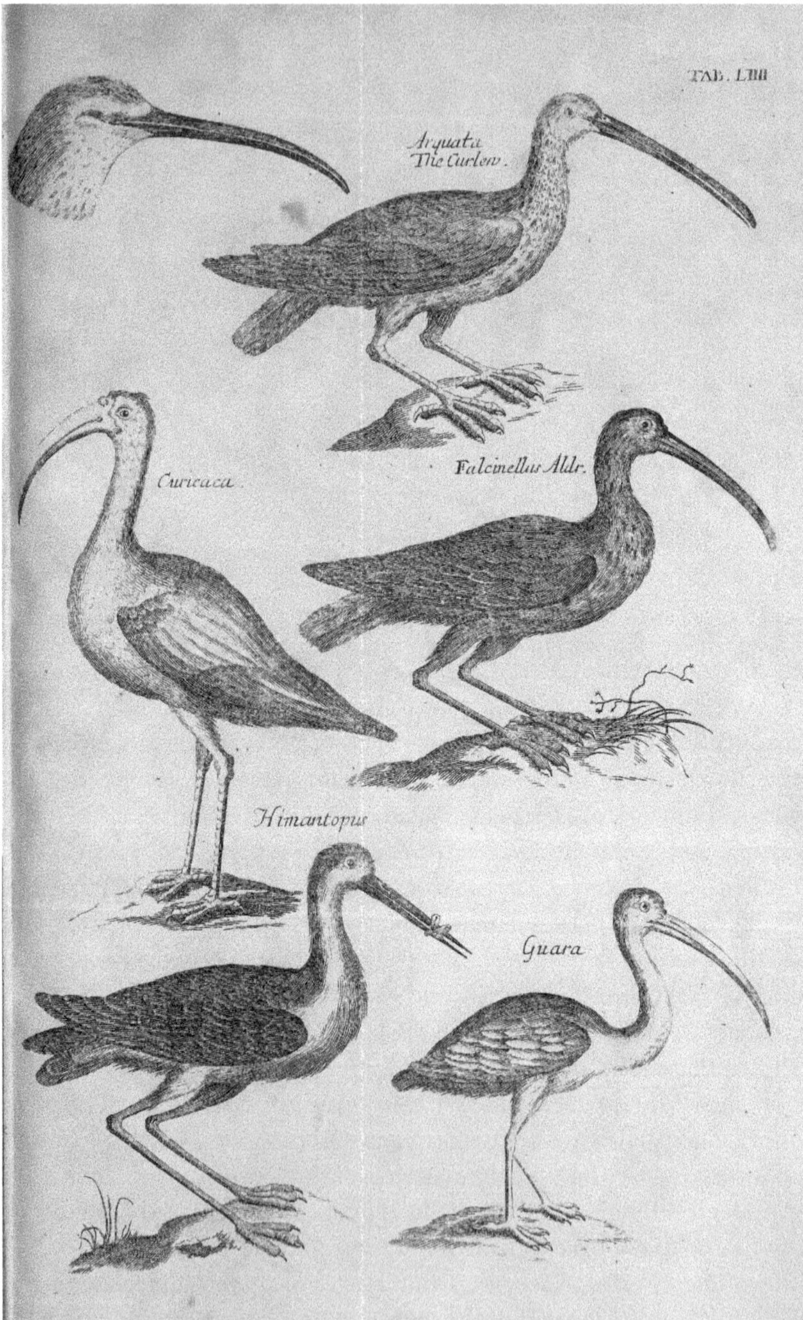

Figure 3.21: Anonymous, *Plate 54*, Engraving, from Francis Willughby and John Ray, *The Ornithology of Francis Willughby* (London: Printed by A.C. for John Martyn, 1678). Cullman Library, Smithsonian Libraries via the Biodiversity Heritage Library.

(Black curve-billed heron). Underneath Willughby has first written "on Alv: 402," which refers to page 402 of volume three of Aldrovandi's *Ornithologiae* where he provides an image of an *Ardea minor alia rostro arcuato* (Lesser Heron, another curved bill).[52] (Fig. 3.19) Another annotation appears to be a reference to Gessner that was never completed: "on Gesner." Willughby, however, has crossed out these annotations and replaced them with "*Falcinellus* Ald: 422." On this page, Aldrovandi discusses the *Falcinellus Avis falcate*, which he illustrates on the facing page. (Fig. 3.20)

While Willughby seems to have settled on identifying this bird based on Aldovandi, Ray did not use this drawing as the basis for the engraving of the *Falcinellus* on plate 54. (Fig. 3.21) Instead, this image is based on the bird shown on page 423 of Aldrovandi. (See Fig. 3.20) The drawing with Willughby's annotations, however, looks like it is in fact a copy of the illustration included in Gessner's *Historia Animalium* of the *Falcinellus*.[53] (Fig. 3.22) The printed image included in Aldrovandi's text was given precedence over both Gessner's image and the drawing that Willughby had in his possession. Studying this drawing and its annotations complicates common assumptions about the usefulness of the drawing for natural history. Instead of depicting an example of this bird that was observed by the draftsman, this drawing appears to be a drawn copy of a hand-colored print that Ray did not accept as being useful for his project. Presented with huge amounts of written and visual information, Ray used the images he judged to be the most useful for their project of digesting the current state of natural historical knowledge.

Exploring the visual source materials for Willughby's *Ornithology* expands our understanding of how accuracy was produced because Ray's heavy reliance on printed materials uncovers instabilities in our understanding of "accurate" as being synonymous with "from the life." The nearness of an image to the life was always mediated, not only by the engraver's burin, but also by acts of collecting. Creating engraved figures "nearly resembling the live BIRDS" then was a process of negotiation between printed images, drawings, and live and dead specimens.

Ray's efforts were appreciated and recognized by his peers and successors. For example, in his catalogue of the Repository of the Royal Society,

52 I have compared the handwriting on this drawing with autograph letters of Willughby's in the Archives of the Royal Society, Early Letters, EL/W3. I want to thank Dorothy Johnston, former Keeper of Manuscripts at the University of Nottingham, for her help in examining Willughby's handwriting.

53 The similarity between this drawing and a set of drawings at the New-York Historical Society is striking. See the "Second Avian Project" discussed in: Olson and Mazzitelli, "Discovery of a Cache," 435–521.

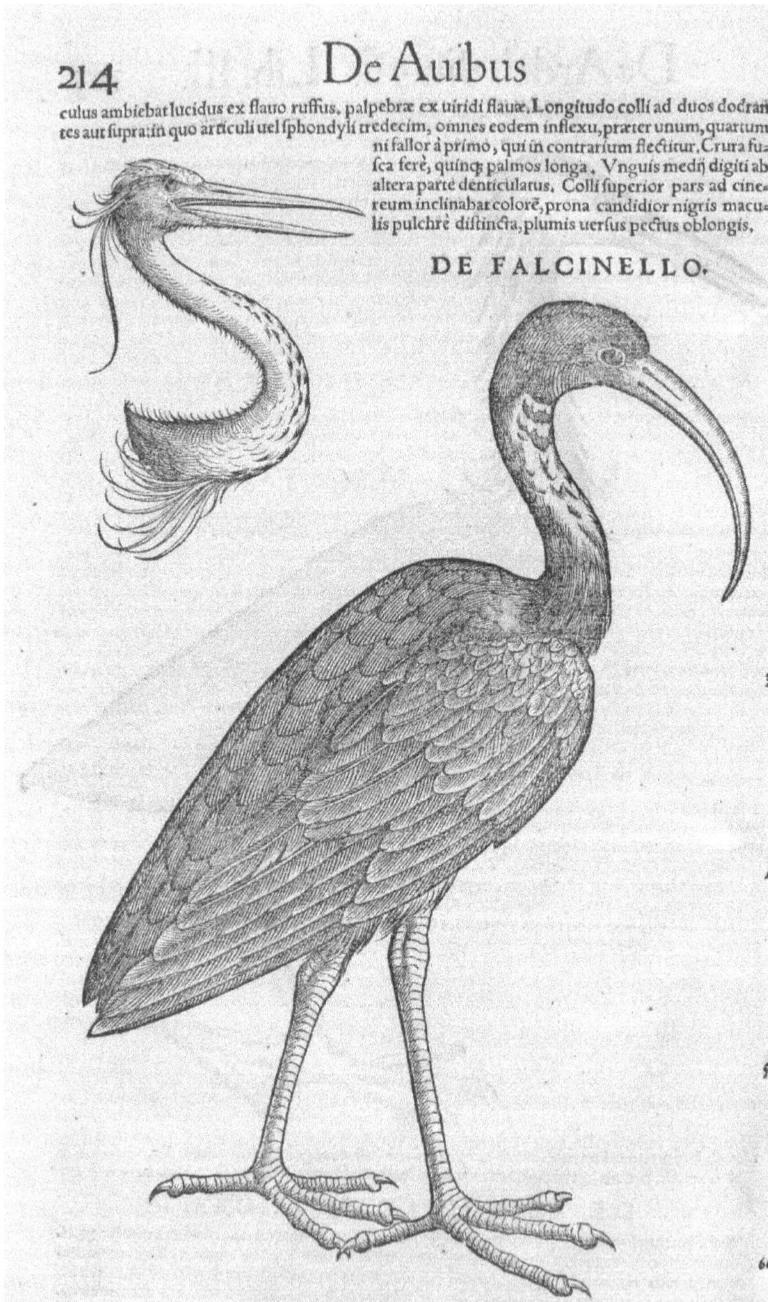
214 De Auibus

culus ambiebat lucidus ex flauo ruffus, palpebræ ex uiridi flauæ. Longitudo colli ad duos dodrantes aut supra: in quo articuli uel sphondyli tredecim, omnes eodem inflexu, præter unum, quartum ni fallor à primo, qui in contrarium flectitur. Crura fusca ferè, quinq; palmos longa. Vnguis medij digiti ab altera parte denticulatus. Colli superior pars ad cinereum inclinabat colorē, prona candidior nigris maculis pulchrè distincta, plumis uersus pectus oblongis.

DE FALCINELLO.

AVIVM

Figure 3.22: Anonymous, *Falcinellus*, page 214, Woodcut, from Konrad Gessner, *Historiæ Animalium Liber III* (Zurich: C. Froschouerum, 1555). Getty Research Institute via the Biodiversity Heritage Library.

Nehemiah Grew refers to Willughby's *Ornithology* multiple times in his descriptions of the birds in the collection and remarks that "...all those Birds, at least, whereof he had the sight, are most curiously and exactly represented."[54] Ray's viewing of many of the birds was understood by his contemporaries to have improved both his descriptions and the plates included in the book. Furthermore, Willughby's *Ornithology* became the standard reference for ornithology well into the eighteenth century. In both his published writings and his manuscript collections, George Edwards (1694–1773) continually referred to Willughby to provide references for his own drawings and descriptions.[55] Edwards was continuing the process of collating observation with the printed record and, again and again, Willughby was the authority. Ray's decisions were viewed as useful by other naturalists both soon after the publication of the work as well as nearly a century after he and Willughby began their project.

Bibliography

I. Manuscript Sources

London, Archives of the Royal Society

EL/W3. Early Letters. Letter of Francis Willughby.

London, British Library

Add. MS 5263. Drawings of Birds in Colours, by G. Edwards, and others, being the originals of his work on the Natural History of Birds. 4pls. 4to. Lond. 1743–50. (154 in number.)

Add. MS 5264. A similar Volume (144 in number.)

Add. MS 5265. A similar Volume (136 in number.)

Add. MS 5266. Drawings of Birds in Colours, crayons, and pencil, by Dr. Browne, and others (163 in number.)

Add. MS 6485. An exact natural Description and Deliniation of the Water-Fowls, Fishes,

Fourfooted Beasts, Insects, and Worms, which are found in the water at Strassburg,

54 Grew, *Musæum Regalis Societatis*, 62.

55 See for instance the drawing of the Rose or Carnation Colored Ouzel in Add. MS 5266/68 and the printed description of it in Edwards, *Natural History of Birds*, vol. 1, 20. This is just one example from the four volumes of drawings of birds that include drawings that were used for Willughby's project as well as Edwards' own drawings. British Library, Add. MS 5263–5266.

Nottingham, University of Nottingham

Mi LM 24. Collection of water-colour drawings and engravings of birds. N.d. mid-17th-early 18th century.

Providence, R.I., Brown University

QL41 .B3 1653 v.1–4. Recht natÿrliche Beschreibung und Abmalung der Wasser Vögel, Fischen, Vierfüssigen Thieren, Insecten, und Gewürm, so beÿ Strassburg in den Wassern gefunden werden, welche ich selber geschossen, gefangen und in mein eigner Hand gehabt.

II. Printed Primary Sources

Aldrovandi, Ulisse. *Ornithologiae, Hoc est De Avibus Historiae Libri XII.* Bologna: Apud Franciscum de Franciscis Senensem, 1599–1603.

Bondt, Jakob de. "Historiæ Naturalis and Medicæ Indiæ Orientalis Libri Sex." In *De Indiæ Utriusque Re Naturali et Medica Libri Quatuordecim*, edited by Willem Piso. Amsterdam: Apud Ludovicum et Danielem Elzevirios, 1658.

Clusius, Carolus. *Exoticorum Libri Decem.* Antwerp: Ex Officinâ Plantinianâ Raphelengii, 1605.

Derham, William. *Select Remains of the Learned John Ray: With His Life.* London: Printed and sold by J. Dodsley and J. Walter, 1760.

Edwards, George. *A Natural History of Birds.* London: Printed for the Author, 1743.

Gessner, Konrad. *Historiæ Animalium Liber III.* Zurich: C. Froschouerum, 1555.

Grew, Nehemiah. *Musaeum Regalis Societatis, or, a Catalogue & Description of the Natural and Artificial Rarities Belonging to the Royal Society.* London: Printed by W. Rawlins, for the Author, 1681.

Hooke, Robert. *Micrographia, or, Some Physiological Descriptions of Minute Bodies Made.* London: Printed by J. Martyn and J. Allestry, 1665.

Nieremberg, Juan Eusebio. *Historia Naturae, Maxime Peregrinae, Libris XVI.* Antwerp: Ex officina Plantiniana Balthasaris Moreti, 1635.

Oldenburg, Henry. "An Account of Some Books." *Philosophical Transactions* 10, no. 120 (27 December 1675): 481–490.

Olina, Giovanni Pietro. *Uccelliera: Overo, Discorso della Natura e Proprieta di Diversi Uccelli.* Rome: A. Fei, 1622.

Piso, Willem. *Historia Naturalis Brasiliae.* Leiden: Apud Franciscum Hackium, 1648.

—. *De Indiæ Utriusque re Naturali et Medica Libri Quatuordecim.* Amsterdam: Apud Ludovicum et Danielem Elzevirios, 1658.

Ray, John. *The Correspondence of John Ray.* Edited by Edwin Lankester. London: Printed for the Ray Society, 1848.

—. *Further Correspondence of John Ray.* Edited by Robert W. T. Gunther. London: Printed for the Ray Society, 1928.

—. *Observations Topographical, Moral, & Physiological Made in a Journey through Part of the Low-Countries, Germany, Italy, and France*. London: Printed for John Martyn, 1673.

Willughby, Francis, and John Ray. *Ornithologiae libri tres* London: Impensis Joannis Martyn, 1676.

—. *The Ornithology of Francis Willughby*. London: Printed by A.C. for John Martyn, 1678.

III. Secondary Sources

Birkhead, Tim, ed. *Virtuoso by Nature: The scientific worlds of Francis Willughby FRS (1635–1672)*. Leiden: Brill, 2016.

Birkhead, Tim et al, "Willughby's Ornithology," in *Virtuoso by Nature: The scientific worlds of Francis Willughby FRS (1635–1672)*. Edited by Tim Birkhead, 268–304. Leiden: Brill, 2016.

Castellani, Carlo. "Aldrovandi, Ulisse." In *Dictionary of Scientific Biography*, edited by Charles Gillispie. New York: Scribner, 1970.

Cook, Harold J. *Matters of Exchange: Commerce, Medicine, and Science in the Dutch Golden Age*. New Haven and London: Yale University Press, 2007.

Duroselle-Melish, Caroline and David A. Lines. "The Library of Ulisse Aldrovandi (†1605): Acquiring and Organizing Books in Sixteenth-Century Bologna." *The Library*, 7th series, vol. 16, no. 2 (June 2015): 133–161.

Fils, Nathan. "Francis Barlow, the King's Birds, and the Ornithology of Francis Willughby and John Ray." *Huntington Library Quarterly*, 78, no. 2 (Summer 2015): 263–300.

Findlen, Paula. *Possessing Nature: Museums, Collecting, and Scientific Culture in Early Modern Italy*. Berkeley: University of California Press, 1994.

Freedberg, David. *The Eye of the Lynx: Galileo, His Friends, and the Beginnings of Modern Natural History*. Chicago: University of Chicago Press, 2002.

Greengrass, Mark et al. "Science of the Move: Francis Willughby's Expeditions," in *Virtuoso by Nature: The scientific worlds of Francis Willughby FRS (1635–1672)*. Edited by Tim Birkhead. Leiden: Brill, 2016, 142–226.

Grindle, Nick. "'No Other Sign or Note Than the Very Order:' Francis Willughby, John Ray and the Importance of Collecting Pictures." *Journal of the History of Collections* 17, no. 1 (2005): 15–22.

Hume, Julian P. "The History of the Dodo *Raphus Cucullatus* and the Penguin of Mauritius." *Historical Biology* 18, no. 2 (2006): 65–89.

Johns, Adrian. *The Nature of the Book: Print and Knowledge in the Making*. Chicago: University of Chicago Press, 1998.

Kusukawa, Sachiko. "*Ad vivum* Images and Knowledge of Nature in Early Modern Europe." In *Ad vivum? Visual Materials and the Vocabulary of Life-Likeness in*

Europe before 1800 edited by Thomas Balfe, Joanna Woodall and Claus Zittel, 89–121. Leiden: Brill, 2019.

—. "The *Historia Piscium* (1686)." *Notes and Records of the Royal Society* 54, no. 2 (2000): 179–197.

Lind, L. R. *Aldrovandi on Chickens: The Ornithology of Ulisse Aldrovandi (1600) Volume II, Book XIV.* Norman: University of Oklahoma Press, 1963.

Lownes, Albert E. "A Collection of Seventeenth-Century Drawings." *The Auk* 57, no. 4 (1940): 532–535.

—. *Renaissance Books of Science: From the Collection of Albert E. Lownes*, edited by David P. Godine and Owen Gingrich. Hanover: Dartmouth College, 1970.

MacGregor, Arthur. "The Cabinet of Curiosities in Seventeenth-Century Britain." In *The Origins of Museums: The Cabinet of Curiosities in Sixteenth- and Seventeenth-Century Europe*, edited by Oliver Impey and Arthur MacGregor, 147–158. Oxford: Clarendon, 1985.

Mandelbrote, Scott. "Ray, John (1627–1705)." In *Oxford Dictionary of National Biography.* Oxford: Oxford University Press, 2005.

Olson, Roberta J. M. and Alexandra Mazzitelli. "The Discovery of a Cache of over 200 Sixteenth-Century Avian Watercolors: A Missing Chapter in the History of Ornithological Illustration." *Master Drawings*, 45, no. 4 (Winter, 2007): 435–521.

Parish, Joylon C. *The Dodo and the Solitaire: A Natural History.* Bloomington: Indiana University Press, 2013.

Parshall, Peter. "*Imago Contrafacta*: Images and Facts in the Northern Renaissance." *Art History* 16, no. 4 (1993): 554–579.

Phillips, John C. "Leonard Baldner, Seventeenth Century Sportsman and Naturalist: An Unrecorded Copy of His Book, Containing His Portrait." *The Auk* 42, no. 3 (1925): 332–341.

Poole, William. "The Willughby Library in the Time of Francis the Naturalist." In *Virtuoso by Nature: The scientific worlds of Francis Willughby FRS (1635–1672)*, edited by Tim Birkhead, 227–243. Leiden: Brill, 2016.

Raven, Charles E. *John Ray, Naturalist, His Life and Works.* Cambridge: The Univerity Press, 1942.

Stresemann, Erwin. *Ornithology from Aristotle to the Present.* Translated by Hans J. and Cathleen Epstein. Cambridge: Harvard University Press, 1975.

Swan, Claudia. "*Ad Vivum, Naer Het Leven*, from the Life: Defining a Mode of Representation." *Word and Image* 11, no. 4 (1995): 353–372.

Thackray, John C. *A Catalogue of Portraits, Paintings, and Sculpture at the Natural History Museum, London.* London: Mansell, 1995.

Wragge-Morley, Alexander. *Aesthetic Science: Representing Nature in the Royal Society of London, 1650–1720.* Chicago: University of Chicago Press, 2020.

4. "These Rude Collections": Accumulating Observations and Experiments

Abstract

This chapter consists of a close study of the illustrated articles related to astronomy and anatomy from the early years of *Philosophical Transactions*. Studying these engravings along with their sources reveals the importance of images to the process of communication within the early modern scientific community and highlights the important role of Oldenburg's network of correspondents in the production of a corporate record of experiments and observations. Moving across languages and among correspondents who rarely, if ever, saw one another's instruments or workspaces, the visual component of these articles was not peripheral to their usefulness, but indeed was a central feature. Studying these articles shows how accuracy was produced through the accumulation of images that circulated among this pan-European community.

Keywords: Henry Oldenburg, *Philosophical Transactions*, Periodicals, Anatomy, Astronomy

In the first issue of the newly organized *Philosophical Transactions,* its editor, Henry Oldenburg, promised in the Introduction to provide his readers "knowledge of what this Kingdom, or other parts of the World, do, from time to time, afford."[1] He clarified his purpose on the title page for the first volume as the full title of the work was: "Philosophical Transactions: Giving some accompt of the present Undertakings, Studies, and Labours of

1 Oldenburg, "Introduction," 1.

Doherty, M.C., *Engraving Accuracy in Early Modern England: Visual Communication at the Early Royal Society*. Taylor & Francis Group, 2022.
DOI 10.5117/9789463721066_CH04

the Ingenious in many Considerable parts of the world."[2] This journal then was to be not just an English effort, but an international one. Oldenburg was at pains to emphasize in his dedicatory letter to the Royal Society that this was not an official publication of the Royal Society, but rather his own private project:[3]

> In these Rude Collections, which are onely the Gleanings of my *private* diversions in broken hours, it may appear, that many Minds and Hands are in many places industriously employed, under Your Countenance and by Your Example, in the pursuit of those Excellent Ends, which belong to Your Heroical Undertakings.[4]

His "Rude Collections" then were the product of his own reading and correspondence with many like-minded individuals scattered across Europe and beyond. Despite Oldenburg's protestations and the evidence Adrian Johns has found for Oldenburg's financial investment in the *Philosophical Transactions*, the pan-European community, nonetheless, perceived the journal as being part of the Royal Society's work.[5] For instance, in an article announcing its commencement in the *Journal des Sçavans*, the *Philosophical Transactions* was introduced to a French audience as being part of the good work of the Society that had the King as its founder.[6] Whether it was seen as Oldenburg's work or that of the Royal Society, the *Philosophical Transactions* participated in a European conversation about advancements in natural knowledge.

In his role as Secretary of the Royal Society, Oldenburg built up a network of correspondents who were interested in pursuing natural

2 The copy of the journal presented to the Royal Society by Oldenburg has an inscription from him dated 30 May 1667. Given the date range printed on the title page, "For Anno 1665, and 1666," it is likely that the title page was printed after the issues for those two years. *Philosophical Transactions* 1, title page.

3 Adrian Johns discusses the private nature of this venture in some detail. Johns, "Miscellaneous Methods," 159–186, at 166. In his article on the early history of the *Philosophical Transactions*, Noah Moxham investigates how the journal survived outside of an institutional context: Moxham, "Fit for Print," 241–260.

4 Oldenburg, "To the Royal Society," n.p.

5 Johns has also noted that the *Philosophical Transactions* was identified with the Royal Society in Europe. "For its pages already represented a public archive of the intellectual propriety of the Society's fellows and correspondents." Johns, *Nature of the Book*, 501.

6 "Car comme la belle Philosophie y fleurit plus qu'en aucun autre lieu de monde; on a pris le soin d'y faire un Journal en Anglois sous le titre de *Philosophical Transactions*, pour fair sçavoir à tout le monde ce qui se découvre de nouveau dans la Philosophie. . .Elle a l'honneur d'avour le Roy d'Angleterre pour Foundateur." "*Philosophical Transactions*," 156.

knowledge, and as he made clear in his Introduction to the first issue, he was constantly encouraging others to join the discussion:

> To the end, that such Productions being clearly and truly communicated, desires after solide and useful knowledge may be further entertained, ingenious Endeavours and Undertakings cherished, and those, addicted to and conversant in such Matters, may be invited and encouraged to search, try, and find out new things, impart their knowledge to one another, and contribute what they can to the Grand Design of improving Natural Knowledge, and perfecting all *Philosophical Arts,* and *Sciences.*[7]

The "addicts" Oldenburg was soliciting in this passage craved further knowledge about the world around them and constantly sought to improve their understanding of phenomena. It was not enough, however, for these "ingenious Endeavours" to merely be pursued independently. Instead, he wanted scholars to share what they discovered to contribute to the Royal Society's larger Baconian program of "improving Natural Knowledge" through the collection and circulation of information. Oldenburg invited his readers to scour the world for "useful knowledge" and encouraged them to openly share their discoveries with one another through his publication.

This chapter explores what it meant for knowledge to be useful and, more specifically, the role of visual evidence in these undertakings. To assert the usefulness of these works, Oldenburg often directly encouraged conversations between his correspondents as he circulated printed and manuscript documents among his network and solicited responses and critiques before an article was published. Moving across languages and among correspondents who rarely, if ever, saw one another's instruments or workspaces, the visual component of these articles was not peripheral to their usefulness, but indeed was a central feature. Studying the publication of articles in the *Philosophical Transactions* shows how a regime of accuracy was produced through the accumulation of images that circulated among this pan-European community.

The *Philosophical Transactions* acted as a clearinghouse and mediator for international debates about improvements in natural philosophy. Oldenburg brought together accounts of experiments, letters between scholars, and translations of texts published on the Continent, circulated them among the Fellows of the Royal Society as well as Continental correspondents, and solicited their responses. Oldenburg's practices as an editor encouraged a

7 Oldenburg, "Introduction," 2.

culture of inquiry that asked how observers and experimenters discerned what was actually being seen, and further, examined how the authors could effectively communicate their observations and results.

With the advent of the *Philosophical Transactions* and the *Journal des Sçavans* in 1665, a new genre of scientific publishing was born.[8] While authors wishing to quickly share a discovery or idea with a larger public had previously published short pamphlets or broadsides, these new journals provided a public space for the research process to be fostered and promoted. The format of the *Philosophical Transactions* was similar to that of political newsbooks, and Noah Moxham has argued that the use of "transactions" in the title was a direct reference to this genre.[9] While the size and the title tied it to political news, the content and engraved illustrations set it apart. By concentrating on illustrated articles, this chapter investigates both the accumulation of knowledge in the pages of the journal and the role of visual epistemologies in that process. The aim of this chapter is to examine the role of the journal in the accumulation of a corpus of knowledge, particularly as it is represented by illustrated articles that study the physical features of Saturn and individual anatomical features in order to build a more complete and thorough understanding of planetary structures and human generation. The serial nature of the journal provided a format for the repeated publication of very similar images and articles that allowed accuracy to develop over time. Accuracy, by this definition, is produced as an additive process, rather than as an established condition, as a selection among prior sources, or as an epistemological object meeting a given set of criteria.

In their astronomical observations and their anatomical experiments, early Fellows of the Royal Society and their correspondents struggled to understand what they saw when looking at their objects of study and had to translate their own visual experiences into images and texts that could circulate in the pages of the journal. The engraver's burin was a crucial instrument for this translation process. Both astronomy and anatomy relied heavily on new

8 For an overview of the early history of both journals see: McCutheon, "'Journal Des Scavans' and the 'Philosophical Transactions'," 626–628; Brown, *Scientific Organizations*, 185–207; Brown, "History and the Learned Journal," 365–378; Andrade, "Birth and Early Days," 9–22; Kronick, *History of Scientific & Technical*; and Kronick, "Notes on the Printing History," 243–268. For other literature on the development of scholarly journals see for instance: Broman, "Criticism and the Circulation of News," 1–26; Bazerman, *Shaping Written Knowledge*; Atkinson, "The *Philosophical Transactions*," 333–371. For an examination of the connections between the illustrated articles in the two journals see: Doherty, "Giving Light to Narrative," 543–569 and Doherty, "'Ordinary skill in Cutts'."

9 Moxham, "Authors, Editors and Newsmongers," 469.

technologies of looking, the telescope and the microscope, which further complicated the question of what each individual saw when looking at a particular object. In the case of astronomy, viewers of the night sky had first to resolve for themselves what they were seeing when looking at Saturn and then clearly convey their observations to Oldenburg, who then included them in the *Philosophical Transactions*. Similarly, with anatomical matters, physicians used the pages of the *Philosophical Transactions* as a space to debate what they saw when they examined dissected body parts—in the cases examined here, testicles—both with the naked eye and with a microscope. In the case of medical knowledge, a more complete understanding of the nature of sexual reproduction developed through the serialized accumulation of experimental evidence with emphasis put on ocular verification of results. By focusing on astronomical and anatomical articles published during Oldenburg's tenure as editor of the journal, this chapter explores the role of both the periodical and the burin in the communication of observations and experimental knowledge. At its core this topic hinges on perception: Oldenburg's practices as an editor encouraged a culture of inquiry that asked how observers discerned what was actually being seen, and further, examined how the authors could effectively communicate their observations and experiences.

The chapter consists of two case studies that illuminate the question of how to communicate one individual's perceptions to a larger group. In the case of astronomy, not only did scholars have to determine what they were seeing through their telescopes, but they also had to reconcile it with Huygens's hypothesis of Saturn's ring, published in 1659, because his theory had come to dominate discussions of the planet.[10] This section looks at how Oldenburg and his correspondents negotiated vexing questions about the nature of the night sky through their images and texts. In the case of astronomical sightings in the *Philosophical Transactions*, the illusion of accuracy was produced through a combination of visual and verbal descriptions in which both the observations and the conditions when the observations were made were carefully delineated.

The second part of this chapter looks specifically at how Henry Oldenburg and the Royal Society promoted a growing interest in unlocking the secrets of generation and how publication in the *Philosophical Transactions* served both as a forum for current research and as a catalyst for further inquiry. The aim of this section is, first, to deepen our understanding of the role played by visualization in the communication of anatomical information and, second, to complicate what it meant to make a useful image. Concentrating on the

10 Huygens, *Systema Saturnium*.

time span between the beginning of the journal in 1665 and the publication of Antoni van Leeuwenhoek's discovery of sperm with his microscope in 1678, not long after Oldenburg's death in September 1677, provides insight into the role of the journal and Oldenburg in fostering debate and discovery.[11]

"The present figure of *Saturn*"

Looking through an early modern telescope and seeing Saturn was not easy.[12] There were a wide range of factors that could hinder an observer's ability to discern what he saw when looking through a telescope. While the quality of the lenses used was of especial concern to instrument makers and observers in the mid-seventeenth century, the air itself also had to be reckoned with as Robert Hooke noted when describing viewing Mars:

> But such had been the ill disposition of the Air for several nights, that from more than 20. Observations of it, which I had made since its being *Retrograde*, I could find nothing of satisfaction, though I often imagin'd, I saw Spots, yet the *Inflective veins* of the Air (if I may so call those parts, which, being interspers'd up and down in it, have a greater or less Refractive power, than the Air next adjoyning, with which they are mixt) did make it so confus'd and glaring, that I could not conclude upon any thing.[13]

While on the one hand establishing his credibility and trustworthiness as a witness, Hooke also tempered expectations about the quality of his own observations. Hooke's attempts to study Mars while it was retrograde were foiled by the poor quality of the London air. While he wanted to believe that he had seen its spot, he could not discount the possibility that his imagination and the atmosphere were playing tricks on him. Satisfactory observations were dependent then on atmospheric conditions as well as the ability to visually discern the differences between artifacts of the conditions of viewing and the phenomena or planetary features under examination. Judgment, then, was key to the production of accuracy in this context.

11 Leeuwenhoek, "Observationes," 1040–1046. Other scholars have noted that the period from the beginning of the journal to Oldenburg's death forms a coherent unit. See for example: Andrade, "Birth and Early Days," 19.

12 An early version of this material greatly benefitted from discussions during the sixth European Spring School on the History of Science and Popularization, "Visual Representations in Science," Menorca, Spain, May 2011.

13 Hooke, "Particulars," 240.

As Albert van Helden has noted, the "problem of Saturn" began as soon as a telescope was pointed toward it by Galileo in 1610.[14] Galileo wrote of his discovery to the secretary of the Grand Duke of Tuscany, his patron:

> ...the star of Saturn is not a single star, but is a composite of three, which almost touch each other, never change or move relative to each other, and are arranged in a row along the zodiac, the middle one being three times larger than the two lateral ones, and they are situated in this form .[15]

Viewed in the context of his recent discovery of the satellites orbiting Jupiter, Galileo saw a planet and two satellites. Van Helden's important research on the first fifty years of observations of Saturn showed how the astronomers experienced a "gestalt switch" in their viewings of the planet.[16] In his consideration of the competing theses put forward to explain Saturn's appearances, Van Helden dates the "eventual acceptance" of Huygens's hypothesis to 1665.[17] This chapter focuses on the period after he contends that Huygens's theory that Saturn is surrounded by a ring was accepted and picks up the story in 1665 with the publication of the first issue of the *Philosophical Transactions* in order to trace the shifts in discussions and images of its ring(s).[18]

From Galileo's first observation of Saturn's "triple-bodied" nature in 1610 to Huygens's realization in 1656 that a ring would "save the appearances" that astronomers had recorded and predict future appearances, astronomers recorded their observations of Saturn with text and images and actively debated their meanings.[19] This combination of text and image allowed readers to participate in the observation process and imagine themselves looking through Galileo's telescope. Huygens, like Galileo before him, used an anagram to announce his ring theory. In a climate of openness laced

14 Albert van Helden has written a great deal on the subject of Saturn in the early modern period. He provides an overview of the history of telescopic observations of Saturn from 1610 to 1665 in two articles: Van Helden, "Saturn and his Anses," 105–121 and Van Helden "'Annulo Cingitur'," 155–174.

15 Galileo Galilei to Belesario Vinta, the secretary of the Grand Duke of Tuscany, in *Opere di Galileo Galilei*, vol. 10, 409–410, quoted in Van Helden, "Saturn and his Anses," 105.

16 Van Helden, "'Annulo Cingitur'," 155.

17 Van Helden, "'Annulo Cingitur'," 155.

18 Following van Helden's lead, I am referring to the ring as singular as it was believed to be so in this period. As will be discussed below, the division in the ring was not known until Cassini "discovered" it in 1675. Van Helden, "'Annulo Cingitur'," note 5.

19 For an overview of the concept of "saving the appearances" in astronomy, see: Lloyd, "Saving the Appearances," 202–222.

with anxiety about priority claims, the anagram played a key role in the communication of discoveries and new theories. In 1659 he revealed the solution to an anagram he had published in 1656: "He is surrounded by a thin flat ring which does not touch him anywhere and is inclined to the ecliptic."[20] As part of his thesis regarding the existence of Saturn's ring, Huygens made a series of predictions about the future appearance of the planet. By 1665 the key elements of his theory had been accepted, what remained to be determined was the angle of inclination of the ring in relation to the planet.

Oldenburg's correspondents were trying to reconcile their own observations with Huygens's theory and their illustrated articles show how they worked to resolve the differences between his theory and what they discovered in practice. In these articles the word "figure" takes on a double meaning that conflates: 1) the appearance of Saturn at a particular moment with 2) the observer's representation of it. The figure of Saturn as drawn by the observer was accepted as being a useful means of communicating "the present figure of *Saturn*."[21] This compression of the two senses of the "figure" obscures the work of the observer and creates the illusion of a perfect translation from viewing to drawing. At the same time, it erases the work of the burin. In effect, the visual and artistic processes associated with constructing the figure are erased when the term "figure" stands as a sufficient basis for the ongoing argument.

To explore the epistemological effects of this erasure, this section focuses on three illustrated articles in the *Philosophical Transactions*, by William Ball, Christiaan Huygens, and Jean-Dominique Cassini, that show, first, the role of images in the circulation of information about Saturn, and second, the ways in which those images were used to navigate dissonances between theory and practice.[22] These articles show that visualizations were central to ongoing discussions of Saturn's form and that there were still concerns about Huygens's theory as further observations did not match Huygens's predictions. Consensus regarding Saturn's ring(s) was developed through an additive process and produced through the accumulation of images and the mobilization of individual's visual judgments.

Saturn was a particularly vexing object for early modern astronomers as improvements in optics revealed previously unknown aspects of the planet's form. The source for the first representation of Saturn in the journal challenges our conceptions of image-making as something that is done by

20 This is Albert Van Helden's translation, Van Helden, "'Annulo Cingitur'," 163.

21 Ball, "Of an Observation," 153.

22 During Oldenburg's tenure as editor, there were nine illustrated articles published on Saturn.

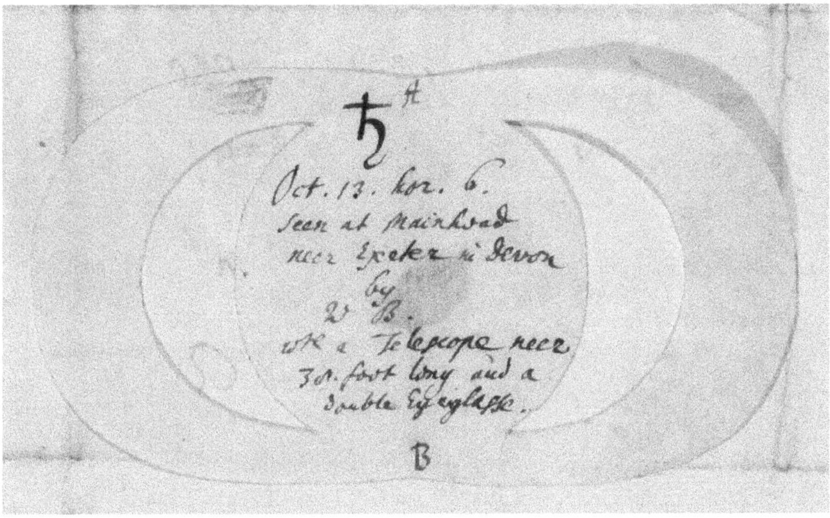

Figure 4.1: Henry Oldenburg, after William Ball, *Saturn*, Paper cutting, 1665. © The Royal Society, EL/B1/108.

making a mark whether on paper or copper. (Fig. 4.1) What William Ball (ca. 1631–1690) sent cannot be called a drawing. Instead, he sent a paper cutting with an account of his viewing of Saturn. This visualization went through a number of hands and multiple versions, that is, it circulated, was revised, and reproduced as it passed through the knowledge network developed by Oldenburg. The version seen by the most people was probably the one included in the ninth issue of the *Philosophical Transactions*.[23] (Fig. 4.2) This visualization highlights the difficulties participants had with representing what they saw and how they struggled to make public what they saw as individuals. Examining this cutting problematizes our conceptions of early modern image making and highlights the care taken by participants to preserve both the material form and the historical context of a visualization as it passed through a network of correspondents and readers.

Although the first report of Saturn's appearance published in the *Philosophical Transactions* is not very long—less than 250 words—it contained a great deal of information about the conditions and content of the observation as well as the circulation of knowledge in the period. Examining how authors framed and illustrated their observations highlights the importance of careful actions to contributions of knowledge that were useful to a community of scholars. These articles, of which Ball's is but one example, call attention to how the authors had to resolve their own observations of Saturn with

23 Ball, "Of an Observation," 152–153.

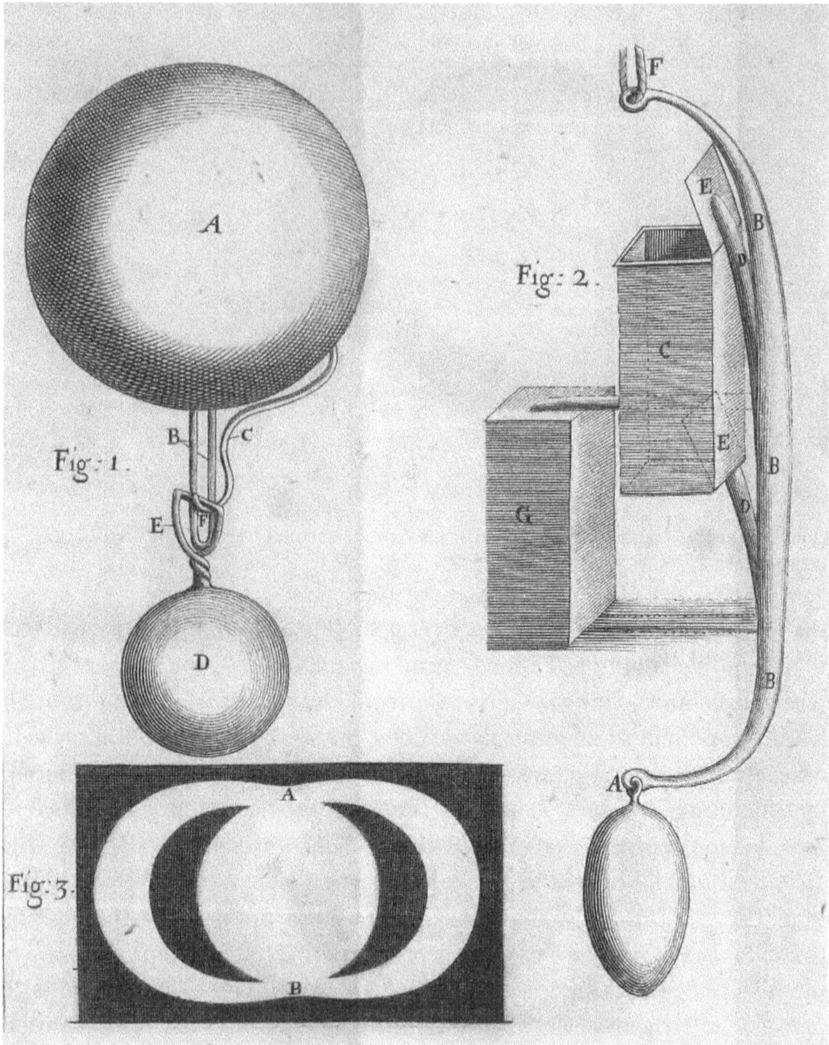

Figure 4.2: Anonymous, *Plate for Issue 9*, Engraving, from *Philosophical Transactions of the Royal Society*, vol. 1, 1665. Dibner Library, Smithsonian Libraries via the Biodiversity Heritage Library.

Huygens's predictions of how it should look at a given time. The pages of the journal provided a space for reconciling theory and observation.

Ball's article, titled "Of an Observation, not long since made in England, of Saturn," began by describing the conditions of his viewing, his apparatus, and the outcomes.

> This Observation was made by Mr. *William Ball*, accompanied by his brother, Dr. *Ball*, *October* 13. 1665. at six of the Clock, at *Mainhead* near

> *Exeter* in *Devonshire*, with a very good *Telescope* near 38 foot long, and a double Eye-glass, as the observer himself takes notice, adding that he never saw that *Planet* more distinct.[24]

This was not the first time Ball turned his telescope toward Saturn, nor his first interaction with Oldenburg and the Royal Society. Rather he was a founding member of the Society and had been making observations of the planet since the 1650s as part of a group of natural philosophers based in Oxford who were concerned with unravelling the problem of Saturn.[25] Furthermore, Ball was often called upon by the Royal Society to participate in the observation of astronomical events, such as eclipses.[26] Although the reader witnessed an observation and not an experiment in this article, the framing of Ball's report fits nicely with Shapin and Schaffer's conception of the technology of virtual witnessing.[27] We know from this introduction who was present at the observation, when and where it took place, and most importantly in the wake of Huygens's published aspersions of others' telescopes, what telescope was used. All of this information served to ground the observation in a particular moment in time, which was important not only in terms of virtual witnessing, but also in terms of understanding how Ball's figure of Saturn fit with Huygens's prediction.

The next sentence in the article introduced Ball's visual evidence of his observation and his uncertainty about his observation's conformity with Huygens's theory:

> The observation is represented by *Figure 3*. concerning which, the Author saith in his letter to a friend, as follows; This appear'd to me the present figure of *Saturn*, somewhat otherwise, than I expected, thinking it would have been decreasing, but I found it full as ever, and a little hollow above and below.[28]

24 Ball, "Of an Observation," 152–153.

25 Van Helden, "'Annulo Cingitur'," 159; Armitage, "William Ball," 167–172; and Gross, "Ball, William."

26 See for example Birch, *History of the Royal Society*, vol. 1, 197 (18 February 1663).

27 Shapin and Schaffer, *Leviathan and the Air-Pump*, 55–60.

28 Ball, "Of an Observation," 153. Not all extant copies of the *Philosophical Transactions* include a plate in issue nine that contains an image of Saturn. Most others have a different plate that is usually labeled in manuscript "No. 9 & 24" which reprinted the instruments from the plate in issue nine. Of the 24 copies I have consulted that have a plate for this issue, seven have the original plate for number nine and 17 have the plate designed for issue number 24. The copy of issue nine at Balliol College, Oxford, includes a drawing of Ball's image as the volume contains the plate that was produced for issue number 24. Both issues nine and 24 have articles concerning

Based on Huygen's ring theory and the predictions he published, Ball had expected Saturn to look a certain way, but he was surprised to find something different. Oldenburg included an engraved figure so that others could study a visual representation of Ball's observations. (See Fig. 4.2) The figure of Saturn here is communicated through a reversed image, that is, the engraving has a black background and the planet with its ring are indicated through blank paper. The resulting image is very flat and provided only the most basic indication of the hollows or indentations that Ball observed. When we look at the visualization that Ball sent along with his account, we begin to understand why this image looks so flat. Instead of putting pen to paper and drawing what he saw, Ball took scissors to paper.

The most obvious and interesting question then is: why a paper-cutting? We might find a preliminary answer by looking at a drawing of Saturn that Ball had sent to Oldenburg with an observation from a few months later. (Fig. 4.3) In April 1666 Ball sent Oldenburg another report of his viewing of Saturn and this time included a small, ink drawing of what he saw.[29] While this drawing does convey a sense of the shape of the ring and how it appeared to be in contact with the planet, it is difficult to resolve exactly what is being shown. Given Ball's rudimentary skills as a draftsman, the paper-cutting was a more effective means of communicating the appearance of the planet not only because it had clearly defined edges, but also because you could pick it up and look at it against a dark background to simulate seeing it in the night sky.[30] There is, however, an underlying assumption in Ball's method, that is by folding the paper in half vertically to cut it, Ball assumed the planet was bilaterally symmetrical.[31] The cutting functioned more like a paper instrument than a drawing. While it does not have moving

"Directions for Seamen, bound for far Voyages." John Couch Adams argued in 1883 that the image of Saturn was left off the issue 24 plate because of an error in it. Adams, "Note on William Ball's Observations," 92–97. This argument does not take into consideration the similarities in the articles published in issues 9 and 24, and there is nothing noted in issue 24 to indicate that an error was found in the plate in issue nine. The articles that are the same in the two issues are: Hooke, "An Appendix to the Directions for Seamen," 147–149; "Directions for Observations and Experiments," 433–42[4]8. This is the only article in issue 24, which also contained material from issue 8, and the figures that were printed in issue 9 (and reprinted in issue 24) are referenced directly in the article.

29 *Correspondence of Henry Oldenburg*, vol. 2, 91. Archives of the Royal Society, Early Letters, EL/B1/108.

30 This is not the only paper cutting in the Royal Society's Archives, for a discussion of a paper cutting of a monstrous herring see: Hunter, *Wicked Intelligence*, 68–70.

31 A version of this material was presented at the Society for the History of Authorship, Reading, and Publishing, Paris, July 2016 and this point was raised during the discussion of my paper.

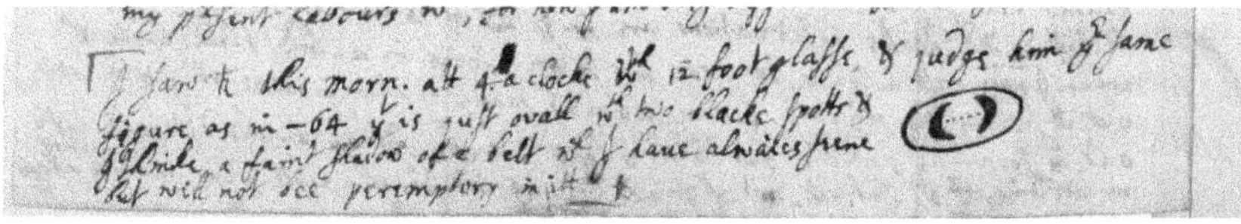

Figure 4.3: William Ball, *Sketch of Saturn*, Ink on paper, 1665. © The Royal Society, EL/B1/108.

parts like a volvelle, it still provided a physical aid to understanding an astronomical phenomenon.

Neither Ball, Sir Robert Moray, who passed the letter on to Oldenburg, nor Oldenburg were content for his observation to be a single, solitary act.[32] Instead, all three were looking to foster a conversation about how Ball's practice of astronomy fit with Huygens's theory. Ball had taken steps towards participating in a conversation about the figure of Saturn by sending his observations to Robert Moray, who in turn sent Ball's report onto Oldenburg and encouraged him to send it on to Huygens: "However I think it worth the Communicating to Mr Hugens, transcribing for him so much of the letter as speakes of these 2. Observations with the description hee makes of his Glass."[33] Moray, then, wanted Oldenburg to send Ball's observations of Jupiter and Saturn along with the description of his telescope, which Ball had included. Ball's figure of the planet was critical to the effective communication of his observation, according to Moray:

> You must make such another figure as is in the letter with the same things writ on it as on the originall: which I would have you send to our president with the letter, and if you cannot cut out another figure to your mind lyke this desire our president to do it. but loose not Mr. Balles letter & figure.[34]

Moray wanted to be sure that Huygens as well as Lord Brouncker, the first President of the Royal Society, were part of the conversation and that they had all of the pertinent visual and verbal information. The safe keeping of the paper cutting and Ball's account were essential to the process.

Although Ball's original letter and cutting are not extant, two of Oldenburg's copies are.[35] Following Moray's suggestion, Oldenburg sent

32 Moray's cover letter is extant and is included in Oldenburg's correspondence. Moray to Oldenburg, 16 November 1665, *Correspondence of Henry Oldenburg*, vol. 2, 608–609.

33 *Correspondence of Henry Oldenburg*, vol. 2, 609.

34 *Correspondence of Henry Oldenburg*, vol. 2, 609.

35 One of Oldenburg's copies of the paper cutting is pasted at the end of a letter from April 1666. Archives of the Royal Society, Early Letters, EL/B1/108. Another is with Huygen's correspondence and is described in his edited correspondence in a footnote to Oldenburg's letter to Huygens.

Huygens an account of Ball's observation, which included a summary of them in French, Ball's original account in English, and the figure recut by Oldenburg.[36] Oldenburg consolidated all of the pertinent information into the representation of Saturn as he inscribed the conditions of the observation onto the center of the planet. All the data Huygens needed to consider Ball's observation was contained within the figure. In his letter to Huygens, Oldenburg asked him to consider Ball's observation in light of his own theory: "You will consider whether it agrees with your theory and tell us, please, what you conclude thereupon."[37] Oldenburg was working to draw Huygens into a conversation about the validity of his own theory in the face of Ball's contradictory evidence prior to his publication of Ball's findings in the *Philosophical Transactions*.

Oldenburg referred to Moray's encouragement to circulate Ball's observations in the article in the *Philosophical Transactions*:

> Whereupon the Person, to whom notice was sent hereof [Moray], examining this shape, hath by Letters desired the worthy Author of the *Systeme of this Planet* [Huygens], that he would now attentively consider the present *Figure* of his *Anses* or *Ring*, to see whether the appearance be to him, as in this Figure, and consequently whether he there meets with nothing, that may make him think, that it is not *one* body of a Circular Figure, that embraces his *Diske*, but two.[38]

It should be noted here that Christopher Wren had earlier circulated a theory that postulated that Saturn was surrounded by an ellipse that touched the planet and that Ball had been an active participant in the observations that led to Wren's theory.[39] Oldenburg was both communicating with Huygens about Ball's findings and letting his readers know that he was circulating this information privately via letter and publicly via the journal. Oldenburg prompted "other Curious men" to join the conversation

"Cette figure est une copie exacte d'un morceau de papier découpé, qui se trouve attaché à la Lettre No. 1502, envoyeé par Oldenburg. L'inscription est de la main d'Oldenburg: évidemment, le papier a été plié en deux, suivant une ligne verticale." *Œuvres Complétes de Christiaan Huygens*, vol. 5, 543–544, n. 2.

36 Oldenburg to Huygens, 23 November 1665, *Correspondence of Henry Oldenburg*, vol. 2, 618–620. See also *Œuvres Complétes de Christiaan Huygens*, vol. 5, 542–544.

37 "Vous considererez, comment elle s'accorde avec vostre Systeme, et nous direz, s'il vous plait, ce que vous aurez conclu là dessus." Original French and translation from *Correspondence of Henry Oldenburg*, vol. 2, 619–620.

38 Ball, "Of an Observation," 153.

39 Van Helden, "Christopher Wren's '*De Corpore Saturni*'," 215, 217.

that he was fostering, and illustrations were central to the comparisons he wanted them to make as he referred directly to the "Notches or Hollownesses, as at A and B."[40] Furthermore, Moray privately continued to encourage Ball to conduct more observations and told Oldenburg that he thought printing his observations would encourage him to keep turning his telescope toward Saturn.[41]

This first account of "the present figure of *Saturn*" in the *Philosophical Transactions* highlights how the journal functioned as a space for a public conversation about reconciling observations to theories. Further, this episode shows how print represents only one portion of a conversation that was going on privately as well. While the paper cutting may be a unique object, the complex circuit of transmission and translation that it traveled is not. As this example shows, in both the public and private aspects of this dialogue, the figure of Saturn was central to the debate.

In 1669 Oldenburg brought Huygens directly into the conversation going on in the pages of the *Philosophical Transactions* by translating and printing an article by Huygens and Jean Picard from the *Journal des Sçavans*.[42] Again the details of the observations were made clear at the beginning of the article:

> An Observation of *Saturne*, made at *Paris*, the *17th* of *August, 1668*. at hor. 11 1/2, at night, by *M. Hugens*, and *M. Picart*, as 'tis describ'd in the *Journal des Scavans* of Febr. 11. 1669. The Observers, imploying a Telescope of 21 Foot...[43]

In this case, the title of the article contained the bulk of the information that located the observation in a particular place and time. As we saw with Ball's observation, a figure was again central to the observation:

> The Observers. . .saw the Planet *Saturn*, as 'tis represented by *Fig*. II. the Globe in the midle manifestly appearing both above and below beyond the Ovale of his Anses; which was hardly discernable the last year.[44] (Fig. 4.4)

40 Ball, "Of an Observation," 153.

41 "I have told him his observation will be communicated to Zulichem [Huygens], and that alone will press him to try it again. the printing will do it much more." Moray to Oldenburg, 15 December 1665, *Correspondence of Henry Oldenburg*, vol. 2, 641.

42 Huygens and Picard, "Observation of Saturne," 900–901. Originally published as: Huygens and Picard, "Observation de Saturne," 11–12.

43 Huygens and Picard, "Observation of Saturne," 900.

44 Huygens and Picard, "Observation of Saturne," 900.

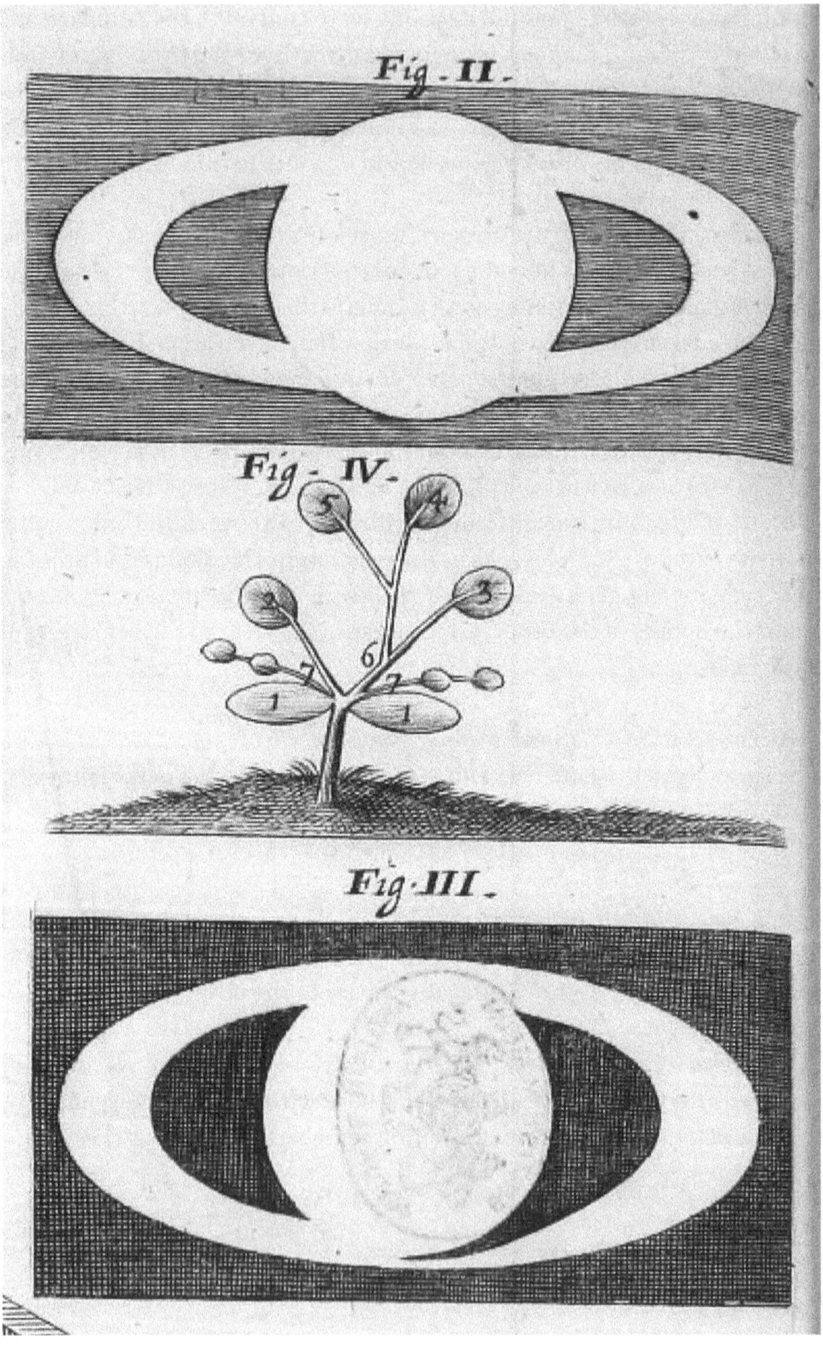

Figure 4.4: Anonymous, after Christiaan Huygens and Jean Picard, *Plate for Issue 45*, Engraving, from *Philosophical Transactions of the Royal Society*, vol. 4, 1669. Natural History Museum Library, London via the Biodiversity Heritage Library.

The framing of this article as it appeared in the *Philosophical Transactions* was very similar to the way it was presented in the *Journal des Scavans*, and the image was re-engraved from it.[45] Oldenburg preserved the overall structure of the article and retained the visual record of Huygens's observation.

The main content of the article consisted of Huygens's revision of his own theory based on his and others' observations. Whereas he had claimed in *Systema Saturnium* that the inclination of the ring to the body of the planet was 23 degrees 30 minutes, he publicly corrected that measurement to 31 degrees in this article.[46] Huygens made space in his theorizing for additional observational evidence and acknowledged that recent observations were "more exact, and made at a more proper time for measuring the obliquity."[47] It is worth noting here that Saturn's anses were not visible when Huygens devised his ring theory.[48] Instead Huygens was working from others' observations, which he compiled in *Systema Saturnium*. The article continued by singling out observations made in 1664 that contributed to Huygens's re-evaluation of the angle of inclination:

> ...that being so, that not onely the Shape, which *Saturn* hath at present, but also all those, that have been noted since the true ones were observed, do perfectly agree with the *hypothesis* of the Ring; and particularly that of 1664. in the beginning of *July**, which was made, and made publick by *Signior Campani*.[49]

This observation of Saturn by Giuseppe Campani (1635–1715) was mentioned in the first issue of the journal in which the quality of his lenses was proved by his observations' conformity with Huygens's theory.[50] In 1665 Campani's publication, *Ragguaglio di nuove Osservationi*, was not available in London and this seems to have still been the case five years later as Oldenburg

45 "Le 17 jour d'Aoust de l'année 1668 à onze heures & demie du soir Mess. Hugens & Picart observerent la Planette de Saturne aven des Lunettes de 21 pieds, & trouverent sa figure telle qu-elle est icy representée, le globe du milieu débordant manifestement par dessus & par dessous hors de l'Ovale de ses anses; ce qui étoit encore à peine visible l'année precedente." Huygens and Picard, "Observation de Saturne," 11.

46 Huygens and Picard, "Observation of Saturne," 900. Van Helden states that this measurement "in fact, erred on the large side, but there would be not further problems in accommodating the disk of the planet in the apparent ellipse of the ring." Van Helden, "'Annulo Cingitur'," 166.

47 Huygens and Picard, "Observation of Saturne," 900.

48 Van Helden, "'Annulo Cingitur'," 155.

49 Huygens and Picard, "Observation of Saturne," 900.

50 "An Accompt of the Improvement of Optick Glasses," 2–3.

provided a different visual reference to gloss Huygens's mention of Campani.[51] Oldenburg added a note after the word "July":

> *See *Fig*. 3. as 'tis to be found in the *French* Letters, written by M. *Auzout* to M. L' *Abbé Charles*, and printed at *Paris*, A. 1665. upon the occasion of the *Ragguaglio de due Nuove Osservationi da Guieseppe Campani*.[52] (See Fig.4.4)

Although Campani's original publication was still not available in London, Oldenburg wanted to ensure that his readers had access to as much visual information as possible so they could judge the quality of Huygens's revision of his theory for themselves.[53] Oldenburg provided readers with as much visual evidence as possible as well as detailed descriptions of his sources so they could judge the validity of Huygens's theory and revision for themselves.

The final illustrated report of Saturn's appearance in the *Philosophical Transactions* that was published during Oldenburg's lifetime provided his community of readers with something that theory could not predict. In 1676 Jean-Dominique Cassini (1625–1712) sent Oldenburg a letter "with a remarkable Observation of Saturn."[54] Whereas Huygens had theorized that Saturn "is surrounded by a thin flat ring," Cassini had found through observation that Saturn's ring was not singular, instead he saw two concentric rings.[55] He wrote further that the inner ring was brighter than the outer.[56] While previous articles, like Ball's, had been concerned with how their observations related to Huygens's theory, Cassini was unconcerned with supporting or refuting the ring theory, as by this point it had been fairly universally accepted and Huygens's clarification of the angle of the ring had placated critics. Instead, he worked from the assumption that it was a ring and realized through observation that the assumption required a more refined account: the ring thesis in fact accounted for *two* rings. (Fig. 4.5)

As was the case with previous accounts of Saturn, Cassini acknowledged that he was participating in a conversation, which Oldenburg was facilitating

51 It seems likely that Oldenburg based this 1665 report on one published in the first issue of the *Journal des Sçavans*. "Ragguaglio de Nuove Osservationi," 3–5.

52 Huygens and Picard, "Observation of Saturne," 900.

53 Based on the digital surrogate in Gallica, the image in Auzout's text is quite different from the one in Campani. Specifically, the negative space between the bottom of the planet and the ring present in the *Philosophical Transactions* image and Auzout's is not present in Campani's image.

54 Cassini, "Extract of Signor Cassini's Letter," 689–690.

55 Cassini, "Extract of Signor Cassini's Letter," 690.

56 "quorum interior exteriori lucidior erat [of which the inner ring was brighter than the outer]." Cassini, "Extract of Signor Cassini's Letter," 690.

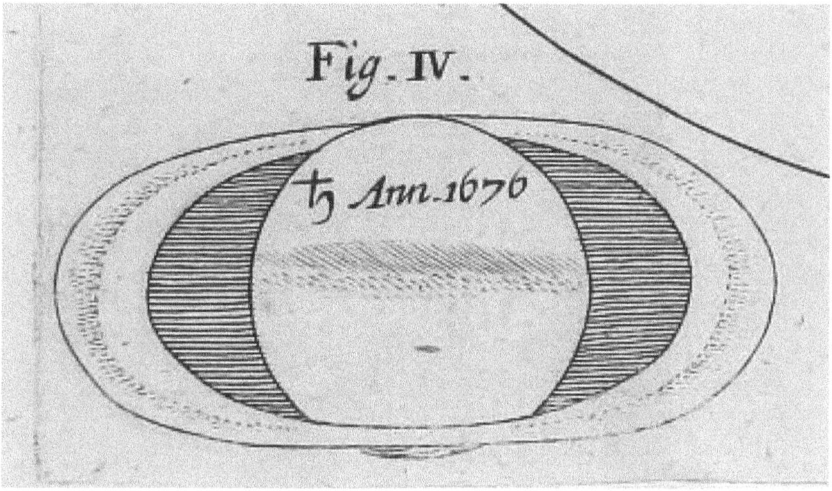

Figure 4.5: Anonymous, after Jean-Dominique Cassini, *Plate for Issue 128*, Engraving, from *Philosophical Transactions of the Royal Society*, vol. 11, 1676. Natural History Museum Library, London via the Biodiversity Heritage Library.

within the pages of the *Philosophical Transactions*. He began his account of this discovery by thanking Oldenburg for sending a copy of Johannes Hevelius's observation of Saturn, which Oldenburg had published in the *Philosophical Transactions* the year before.[57] Cassini also participated in the conversation about the quality of instrumentation and authority of witness conditions. He used Hevelius's illustration of Saturn included in the *Philosophical Transactions* as evidence of the poor quality of Hevelius's telescope, noting that "I see from the representation of Saturn as observed by the distinguished Hevelius a year ago that he employed telescopes very inferior to ours."[58] (Fig. 4.6)

Throughout the early years of the *Philosophical Transactions*, Hevelius had been a constant part of the conversation about "the present figure of *Saturn*." Of the nine illustrated accounts of Saturn published during Oldenburg's tenure as editor, three had been sent by Hevelius.[59] As Mary Winkler and

57 Hevelius included his most recent observation of Saturn at the end of an account of a solar eclipse. "So far the Learn'd *Hevelius*; who was also pleased to communicate his Observation of the Figure of *Saturn*, as it appeared to him in *August*, 1675. to be seen in *Tab.* I. Fig. 3." Hevelius, "Eclipsis Solis Anno 1675," 661.

58 "Ex schemate *Saturni* à Clasissimo *Hevelio* ante annum observato video, cum Telescopiis, nostris longè inferioribus, uti." Cassini, "Extract of Signor Cassini's Letter," 690. The translation is from *Correspondence of Henry Oldenburg*, vol. 13, 39.

59 In addition to the one cited above from 1676, Oldenburg published two other accounts of Hevelius's observations: Hevelius, "Extract of a Letter of M. Hevelius," 2087–2091; Hevelius, "Extract of two Letters of M. Hevelius," 3027–3033.

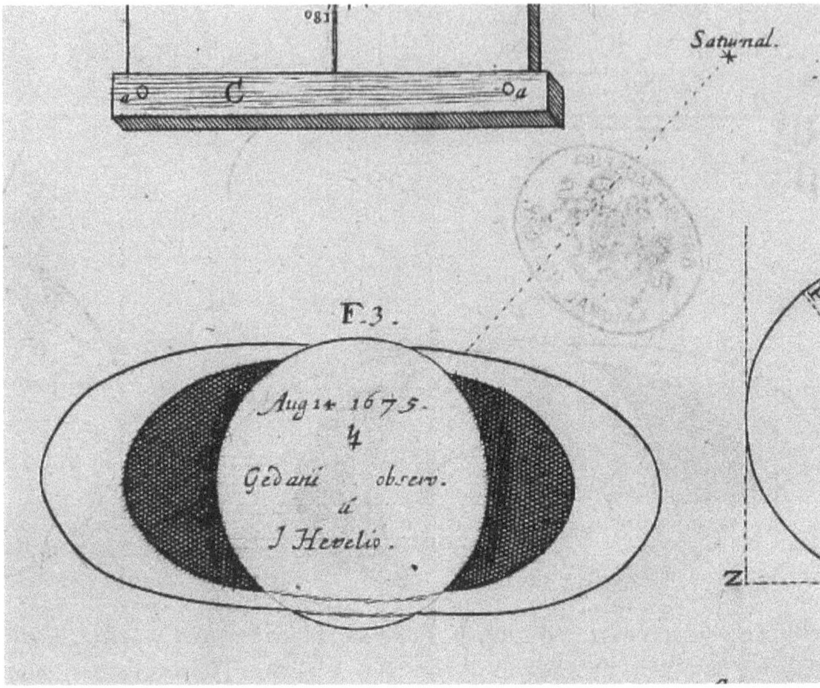

Figure 4.6: Anonymous, after Johannes Hevelius, *Plate for Issue 127*, Engraving, from *Philosophical Transactions of the Royal Society*, vol. 11, 1676. Natural History Museum Library, London via the Biodiversity Heritage Library.

Albert Van Helden have argued, Hevelius was "the acknowledged authority on telescopic astronomy," and as John Evelyn noted, his image making skills were exemplary for other natural philosophers.[60] All this is to say then that Cassini's charge against the quality of Hevelius's telescope and the resulting image should not be taken lightly. The blame cannot be put at the feet of the engraver either as the image in the *Philosophical Transactions* looks remarkably similar to the drawing that Hevelius sent to Oldenburg. (Fig. 4.7)

However, Cassini's image did not provide irrefutable proof of the quality of his optics or his observations. He ended his letter by apologizing for the enclosed figure because it was "somewhat rudely drawn with a hasty pen."[61] The letter Cassini sent to Oldenburg relating his observations of

60 Winkler and Van Helden, "Johannes Hevelius," 98.

61 "Eius delineationem, utcunque rudem, properante calamo hic adieci." *Correspondence of Henry Oldenburg*, vol. 13, 39. The Halls noted that the drawing is no longer with the letter and instead reproduce the image from the *Philosophical Transactions*. It is now located in the volume of the Classified Papers on Astronomy in the Archives of the Royal Society, CLP/8i/32.

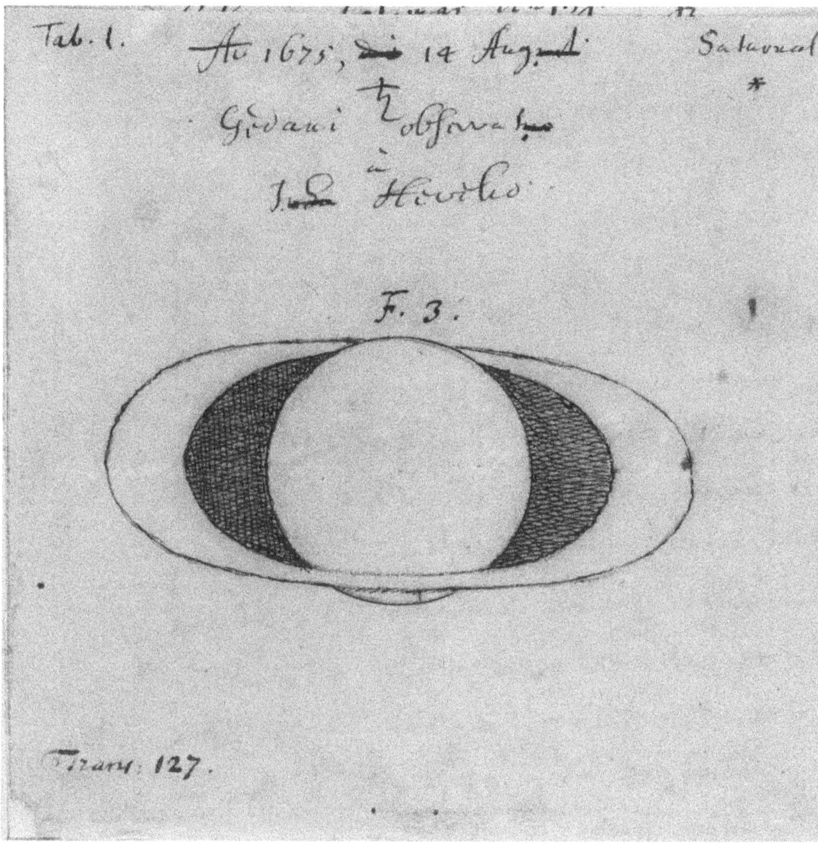

Figure 4.7: Johannes Hevelius, *Sketch of Saturn*, Ink on paper, 1675. © The Royal Society, EL/H2/42.

sunspots and Saturn included three drawings: two of sunspots and one of Saturn. (Fig. 4.8) The shaky ink drawing of Saturn clearly shows the division he saw between the inner and outer rings even if the edge of the division wanders slightly on the left-hand side of the drawing. The brightness of the inner ring is indicated through unadulterated paper, while the darker outer ring has a mottled appearance. The engraver has regularized the shape of the ring but has maintained the visual distinction between the textures of the inner and outer rings. (See Fig. 4.5) The burin worked to standardize Cassini's observation and put them into a vocabulary that was familiar to readers of the *Philosophical Transactions*. What Cassini offered then was not a clearer figure of Saturn but rather a clear understanding of "the present figure of *Saturn*." Accuracy was not produced solely through images. Instead text and image worked together to create the illusion of accuracy.

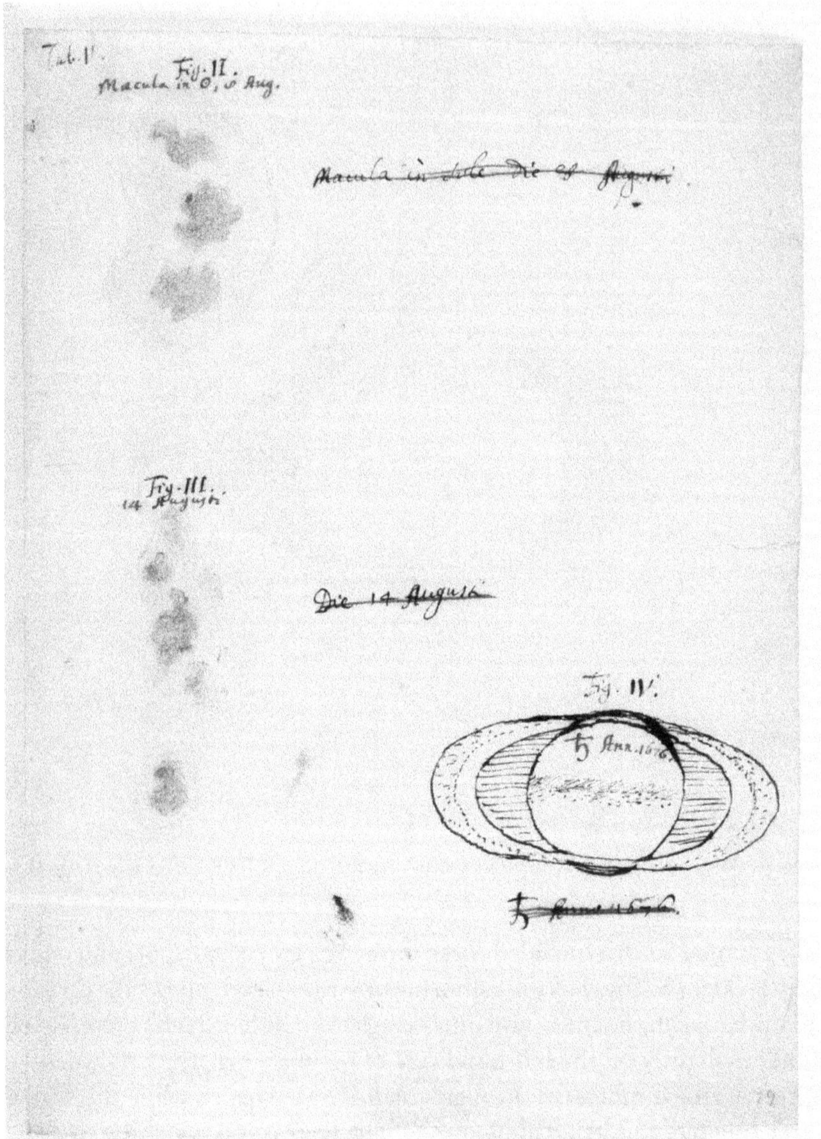

Figure 4.8: Jean-Dominique Cassini, *Sketch of Sunspots and Saturn*, Ink on paper, 1676. © The Royal Society, CLP/8i/32.

The problem of Saturn was finally resolved through a combination of observing the planet under different circumstances and, subsequently, weighing those observations against Huygens's theory. It was only after observational data had helped refine Huygens's theory that Cassini was able to look at the single ring and realize it has in fact multiple rings. The *Philosophical Transactions* served as a space for the circulation of new observations as well as a forum for debating the relationship between theory and practice. Taken together, these articles about Saturn's ring(s) highlight the complex relationship between theory and practice in seventeenth-century astronomy. Neither theoretical modeling nor the practical recording of observations alone was going to reconcile astronomers' representation of Saturn on paper and through the lens to its actual presence in the sky and, finally, in astronomers' minds. In place of thinking about accuracy in this case as a static term, its development is represented in the necessary dialogue between theory and practice, in which the *Philosophical Transactions* served as a space for that discussion and negotiation. Oldenburg's editorial tactics created the illusion of a conversation going on among the men who were writing to him and whose work he was reading, collecting, and recirculating. Although from a modern perspective it is easy to view Huygens's publication of *Systema Saturnium* in 1659 as the moment it became universally known and accepted that Saturn is surrounded by a ring, actually it took years for the problem to be solved, and the articles Oldenburg included in the *Philosophical Transactions* were an important part of the process. Discussions about the figure of Saturn were tied up with larger debates about the place of theorization in the study of nature, and visual evidence was central to these debates in astronomy. The problem of Saturn could only be resolved with the help of images, and authors incorporated images within the text of their observations to establish the usefulness and accuracy of their research habits and images. The engraver's burin was essential to this process as it allowed a disparate community to see the figure of Saturn.

"With so much care and exactness"

Where astronomers were concerned in their observations of Saturn with reconciling their individual viewings of the planet with Huygens' theory and predictions, in the case of anatomical research into reproduction there were competing interpretations of similar sets of data. The key voices in these conversations included physician-fellows, such as Timothy Clarke (d. 1672) and Edmund King (1630–1709), and others working on the Continent, including Regnier de Graaf (1641–1673) and Antoni van Leeuwenhoek (1632–1723). Comparative methodologies were key to the success of the accumulation of anatomical knowledge both in terms of chronology and cross-species studies. Focusing on issues related to "generation" creates a second case study for examining the role of the *Philosophical Transactions* in encouraging intellectual development early in its history. A. Rupert Hall has written briefly about the role of the Royal Society in the debates surrounding mammalian generation and concluded: "It must be said that the role of the Fellows of the Royal Society in this dispute was not an admirable one: they displayed impatience, complacency, and an excessive fondness for *verba* in place of *acta*."[62] While Hall spent little time on this unflattering episode in the history of the Royal Society, it is a productive moment for examining the role of visual judgment in the production of experimental knowledge.

What sets this group of articles apart from those concerned with Saturn is the concomitant development of research methods and results. Although there were debates about the quality of individual's telescopes, *how* to look at Saturn was not open for debate. In contrast, how to prepare anatomical specimens was. Oldenburg's "rude collections" provided a forum for the accumulation of methodological and experimental innovation. This section explores three types of accumulation: first, the accumulation of references to experiments; second, the accumulation of revisions that sought to distinguish new evidence from previously published evidence; and third, the accumulation of discoveries that fundamentally changed scholars' understandings of the topic. All three processes are dependent on the serialized format of the journal and are caught up in, rather than separate from, manuscript and verbal exchanges of information. Throughout these articles, images were key to communicating both what was new and what was important, and the authors stressed the need for visual evidence in their arguments.

Not all the illustrated anatomical or medical articles in the *Philosophical Transactions* during this early period were concerned strictly with questions of generation. Rather these articles fall into two general categories: those that accumulated histories of unusual occurrences, be they monstrous births or abnormal growths, and those that pursued a research question.[63] Hall argued that Oldenburg viewed the collection of abnormalities as part of the Royal Society's Baconian project.[64] Both types of articles capitalized on the serialized format of the journal as they slowly built up a repository of information and often referred back to previous articles for the sake of comparison. It is these processes of accumulation and comparison that I examine, particularly in the context of the second type of articles.

62 Hall, "Medicine and the Royal Society," 436.

63 For instance: Durston, "Narrative of a Monstrous Birth," 2096–2098; Anonymous, "Account Concerning a Woman," 969–970; and King and de Graaf, "Some Observations," [1043]–1047.

64 Hall, "Medicine and the Royal Society," 429.

In 1668 two illustrated articles related to human generation were published in the *Philosophical Transactions.* These articles highlight two of the related, but different forms of experimental evidence accumulated in the journal in its early years. The impetus for the publication of these articles was Regnier de Graaf and Johannes van Horne's publications of their findings regarding mammalian generation, which were reviewed in the *Philosophical Transactions* in April 1668.[65] The first article by Timothy Clarke appeared in the next issue and presented an account of experiments conducted over nearly ten years often at the behest of or at the very least with prodding from the Royal Society.[66] The second, a reprint of a hard to find broadside, made previous, yet hard to access, research available to the audience of the *Philosophical Transactions*.[67] Taken together these articles made two types of previously inaccessible information available, whether due to its private nature or its scarcity, to the audience of the *Philosophical Transactions.* Through these two articles, published within a few months of one another, both first-hand and previously published experimental knowledge accumulated in the pages of the journal.

Timothy Clarke was a founding member of the Royal Society and one of King Charles's physicians.[68] Throughout the early meetings of the Royal Society, Clarke was among a group of physician-Fellows who were often charged with pursuing medical or anatomical research in order to present it at meetings of the Society.[69] As with the range of medical articles published in the *Philosophical Transactions,* Clarke's activities spanned from experiments related to generation and blood transfusions to accounts of medical abnormalities. One of the earliest mentions of Clarke outside of lists of meeting attendees in Birch's *History of the Royal Society,* is a reference to experiments related to claims concerning spontaneous generation in the case of vipers.[70] Clarke was also involved in the examination of medical abnormalities as in 1663 when he was sent by the Royal Society to interview "a woman, who has carried her child eighteen years in her belly, and then excluded the bones

65 "Account of Some Books," 663–604[664].

66 Clarke, "A Letter, written to the Publisher," [672]–682.

67 Aubry, "Testis Examinatus." 843–844.

68 Goodwin, "Clarke, Timothy." Birch included him in the group of men whose meetings in Oxford presaged the founding of the Royal Society, and Clarke was also listed among the founding Fellows. Birch, *History of the Royal Society,* vol. 1, 3–4.

69 See for instance Birch, *History of the Royal Society,* vol. 1, 25 (injections into veins); vol. 1, 67 (dissection of a dog); vol. 1, 77 (dissection of rabbits).

70 Birch, *History of the Royal Society,* vol. 1, 22. Clarke was charged with bringing in a "relation of the production of young vipers from the powder of the liver and lungs of vipers." Robert Boyle and Clarke were then asked "to procure an history of vipers."

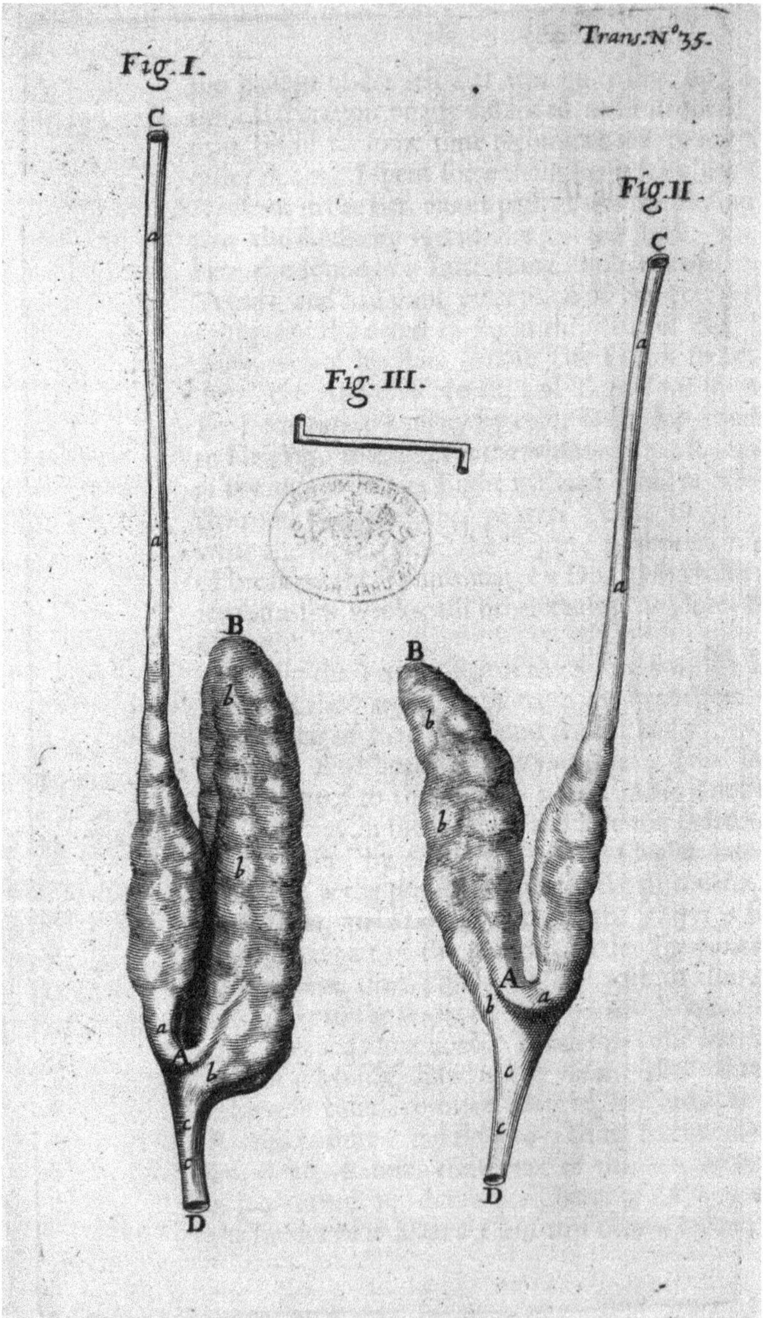

Figure 4.9: Anonymous, after Timothy Clarke, *Plate for Issue 35*, Engraving, from *Philosophical Transactions of the Royal Society*, vol. 3, 1668. Natural History Museum Library, London via the Biodiversity Heritage Library.

of it, after an extraordinary manner, by an abscess in the side of her belly."[71] The variety of Clarke's interests was fairly typical of the medical doctors who were Fellows of the Royal Society and of the larger membership.[72]

After much prodding, Clarke finally prepared for publication the results of a series of experiments performed at Society meetings and in private with other fellows. Clarke's rather lengthy article recounted his research over the previous decade while constantly insisting on how his ideas differed from others and how various Englishmen had been the first to make discoveries regarding "the Origin of Injection into the Veins, the Transfusion of Bloud, and the Parts of Generation."[73] The images he included were of two examples of vas deferens and seminal vesicles from humans that he and Richard Lower had dissected and a small image of a vein related to his experiments with blood transfusion. (Fig. 4.9) The overall tone of the article was quite contentious, and its publication furthered his disagreements with other scholars, especially Regnier de Graaf, whose priority for the discovery of the vas deferens and the seminal vesicle Clarke contested. Publication in the *Philosophical Transactions* provided a wider audience than the membership of the Royal Society and their correspondents with an account of experiments conducted by Fellows regarding "the parts of generation." This article both furthered debate and research and provided a published record that could be referred to in subsequent additions to the literature.

While Clarke's article was a first-hand account of experiments, in the December 14, 1668 issue of the *Philosophical Transactions,* Oldenburg included a complete reprint of a Latin broadside that had been published ten years earlier, which recirculated previously published experimental evidence.[74] (Fig. 4.10) He stated his reasons clearly in the header to the article:

> Testis Examinatus. This is the Title of a printed Page, formerly (viz, Anno 1658) at Florence, by Vauclius Dathirius Bonglarus, and now, by reason of the great scarcity of the Original here, desired to be inserted in this Tract: which is the rather done at this time, because the subject therein consider'd, is now under a severer Examination than ever, amongst the Curiouser Anatomists both here, in France, and Holland.[75]

71 Birch, *History of the Royal Society,* vol. 1, 256.

72 For an overview of the medical interests of the early Royal Society see: Hall, "Medicine and the Royal Society," 421–452.

73 Clarke, "A Letter, written to the Publisher," 672. The entire article, which was published in Latin, is translated in *Correspondence of Henry Oldenburg,* vol. 4, 359–369.

74 Aubry, "Testis Examinatus," 843–844; Bonglaro [Aubry], *Testis Examinatus.*

75 Aubry, "Testis Examinatus," 843.

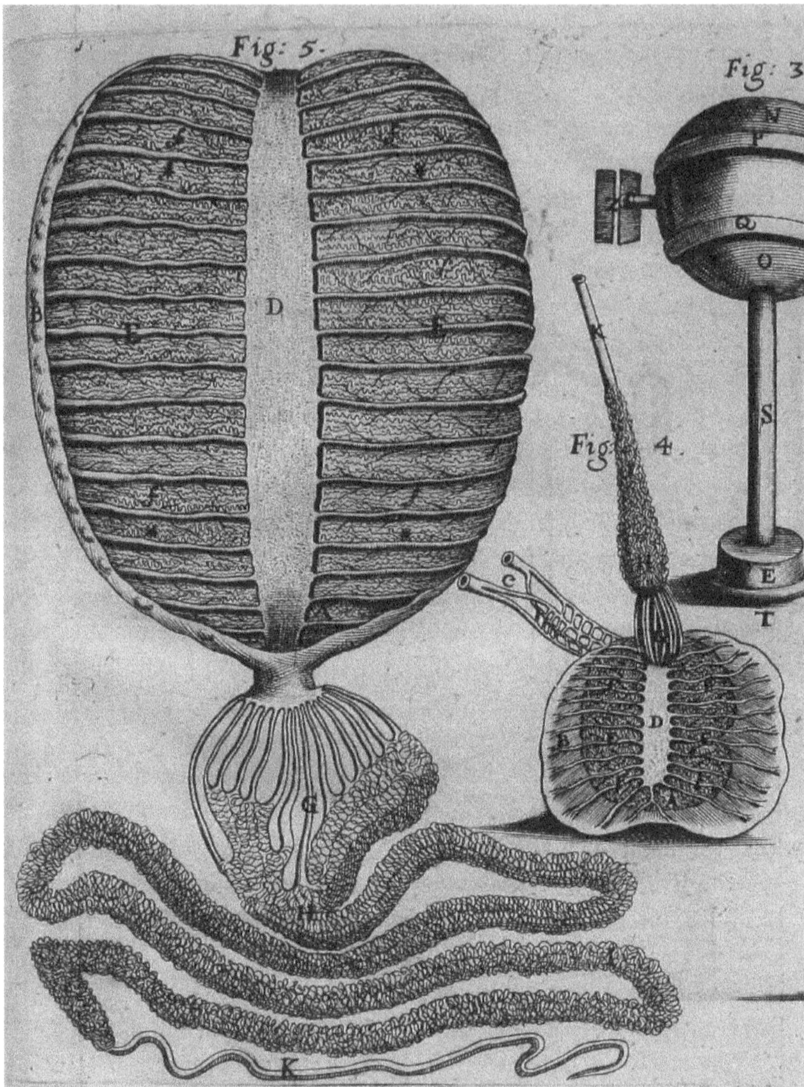

Figure 4.10: Anonymous, after Claude Aubry, *Plate for Issue 42*, Engraving, from *Philosophical Transactions of the Royal Society*, vol. 3, 1668. Natural History Museum Library, London via the Biodiversity Heritage Library.

Vauclius Dathirius Bonglarus is an anagram for Claudius Aubrius Lotharingus, who was also known as Claude Aubry.[76] At the time Aubry published this broadside, he was lecturer of anatomy at the University of Pisa.[77] Figures

76 Cornell University Library Catalogue and Placciu, *De Scriptis & Scriptoribus*, 168.

77 Conforti, "The Experimenters' Anatomy," 34.

four and five in the plate in the *Philosophical Transactions*, which were copied from Aubry's broadside, provided visual evidence of his, self-described, careful consideration, patient dissections, and curious looking.[78] As with previous examples discussed, Oldenburg framed his editorial decisions in such a way that the *Philosophical Transactions* were positioned as a forum for the curious to discuss their findings.

Similarly, Aubry ended his broadside by encouraging others to pursue a series of questions he saw as needing further research.[79] Despite trying to remain neutral in other contexts, Oldenburg inserted a footnote to Aubry's last sentence in English that weighted the intellectual scales toward English contributions:

> Besides what *D. de Graef* hath since publisht upon this Subject, the same hath been lately examin'd by the *R. Society* with so much care and exactness, that now there remains but little doubt of what is conceived, and *hath been* so many years agoe, by able Anatomists here in *England*, of the structure of the *Testicles, viz*. that they are a *Congeries*, or heap of very fine vessels, that may be drawn out like thred, and distinctly expos'd to the Eye.[80]

Oldenburg praised his colleagues "care and exactness" and used that as part of his rationale for accepting their findings. With this remark, Oldenburg seemingly closed the debate on the structure of the testicles and gave weight to Clarke's disputing of De Graaf's claims for priority in favor of the "able Anatomists here in *England*." However, as we will see, the debate was far from over.

Clarke's accusations and priority claims, on behalf of George Joyliffe (1621–1658) and Thomas Wharton (1614–1673), led to further experimentation at the Royal Society and a flurry of correspondence between Oldenburg and de Graaf.[81] Because his methods and results were questioned, primarily by Clarke, a group of physician-Fellows were given the task in October 1668 of recreating de Graaf's findings concerning testicles.[82] Two weeks after receiving their charge, the group, which now included Clarke and Oldenburg, reported that they "did not make out what Dr. de Graaf had asserted of the

78 Aubry, "Testis Examinatus," 844.

79 "Alia si noueris hac in parte quæ meos latuerint oculos, euulges enixe rogo, meo aliorumque commodo." Aubry, *Testis Examinatus*.

80 Aubry, "Testis Examinatus," 844.

81 With regard to the priority claims see: Hall, "Medicine and the Royal Society," 435.

82 Edmund King, Peter Balle, and Thomas Allen were charged with recreating de Graaf's experiments with testicles. Birch, *History of the Royal Society*, vol. 2, 317.

substance of the testicles, though they neither refuted it…it was moved, that more experiments might be made about it."[83] Finally, the next week, Edmund King "related, that he had prosecuted the experiment about the substance of the testicles, and shewed the society some of those of a buck-rabbet, which seemed indeed to several of the members to be made up of vessels…"[84] Even so, Clarke continued to contest both the results and de Graaf's claims of discovery.[85] Six weeks after Fellows started their concentrated attempts to recreate de Graaf's results, Edmund King "brought in a human testicle unravelled, and thus glued on a glass."[86] While the majority of the Fellows seem to have been satisfied with King's results, Clarke's dispute with de Graaf continued and King's results remained unpublished.

The event that finally led to the publication of King's refinement of de Graaf's experimental method was an exchange of specimens. At the October 21, 1669 meeting of the Royal Society, Oldenburg presented to the Fellows "several letters and other papers with some curiosities," which he had received during their summer break.[87] Included in this group was "a little glass sent to Mr. Oldenburg from Dr. de Graaff, containing a *testiculus gliris* unraveled and swimming in spirit of wine, designed to prove, that that organ is made up of nothing but small vessels."[88] De Graaf clearly stated his motivations for sending the specimen in an accompanying letter: "To corroborate my words with deeds I am sending to you, Mr. Oldenburg, the testicle of a dormouse, unraveled by my method, so that you may see whether such glands as Mr. Clarke proposes in his letter…are to be found in the testes."[89] Throughout his account De Graaf stressed his belief in the persuasive role of the visual in this debate.

Oldenburg seems to have agreed as he published both an image of King's prepared specimen and de Graaf's.[90] (Fig. 4.11) In his report, King emphasized the importance of visual evidence and his reliance on others such as Robert Hooke, to confirm his interpretation of what he saw both with the naked eye and "by the help of a good Glas."[91] As this was a particularly visual debate,

83 Birch, *History of the Royal Society*, vol. 2, 321.
84 Birch, *History of the Royal Society*, vol. 2, 327.
85 Birch, *History of the Royal Society*, vol. 2, 328.
86 Birch, *History of the Royal Society*, vol. 2, 333. A week later King submitted a written account of his experiments. Birch, *History of the Royal Society*, vol. 2, 335.
87 Birch, *History of the Royal Society*, vol. 2, 396.
88 Birch, *History of the Royal Society*, vol. 2, 397.
89 *Correspondence of Henry Oldenburg*, vol. 6, 122.
90 King and de Graaf, "Some Observations," [1043]–1047.
91 King and de Graaf, "Some Observations," 1403[1043].

King went to great pains to confirm his own individual viewing of the specimen with others, to the end that they "seem'd to be fully satisfied."[92] The prepared specimen served to consolidate these repeated viewings of multiple individuals into one object that many could examine, such that the composition of the testes "will evidently appear to the bare Eye."[93] King went to great lengths in his report to describe the various ways he went about preparing his specimens as well as describing the variety of animals he examined. His ultimate goal in his comparative studies was to understand human testicles: "the exact knowledge of whose fabrick we suppose to be chiefly aimed at in this kind of Inquiry."[94] King believed his visual evidence along with a detailed explanation of his methods could show others that he had uncovered "the true genuine substance *Testiculi humani*."[95] Throughout the debate between Clarke and de Graaf, the *Philosophical Transactions* served to both publicize de Graaf's findings regarding the make-up of testicles and promote the journal as a forum for debate.

While the articles discussed thus far worked to accumulate a record of experiments and refinements of evidence and methods, the final article I will discuss was the first published account of a discovery. Antoni van Leeuwenhoek was first introduced to Oldenburg in a letter from de Graaf in 1673 and soon after began corresponding directly with Oldenburg about his observations and his microscopes.[96] Over the course of the next fifty years, until his death in 1723, Leeuwenhoek sent hundreds of letters to the Royal Society.[97] Although the first report sent by De Graaf did not contain images, Oldenburg asked that future ones would as the Fellows "were extremely anxious to be able to examine figures of the triple sting observed in the bee and of the limbs noted in the same insect."[98] In his first response to De Graaf, Oldenburg also began to shape and direct Leeuwenhoek's researches and, perhaps not surprisingly, issues of generation were first on Oldenburg's

92 King and de Graaf, "Some Observations," 1403[1043].
93 King and de Graaf, "Some Observations," 1044.
94 King and de Graaf, "Some Observations," 1044.
95 King and de Graaf, "Some Observations," 1044.
96 *Correspondence of Henry Oldenburg*, vol. 9, 602–603. For an overview of their correspondence and Leeuwenhoek's contributions to microscopy see: Ruestow, *Microscope in the Dutch Republic*, chap. 6 (Leeuwenhoek I: A clever burgher) and chap. 7 (Leeuwenhoek II: Images and Ideas).
97 The letters in the Archives of the Royal Society from Leeuwenhoek fill four volumes of the Early Letters series and contain more than 300 items, including letters and their translations (EL/L1–L4). Ruestow notes that there are 116 articles in the *Philosophical Transactions* that are drawn from Leeuwenhoek correspondence. Ruestow, *Microscope in the Dutch Republic*, 150.
98 *Correspondence of Henry Oldenburg*, vol. 9, 654.

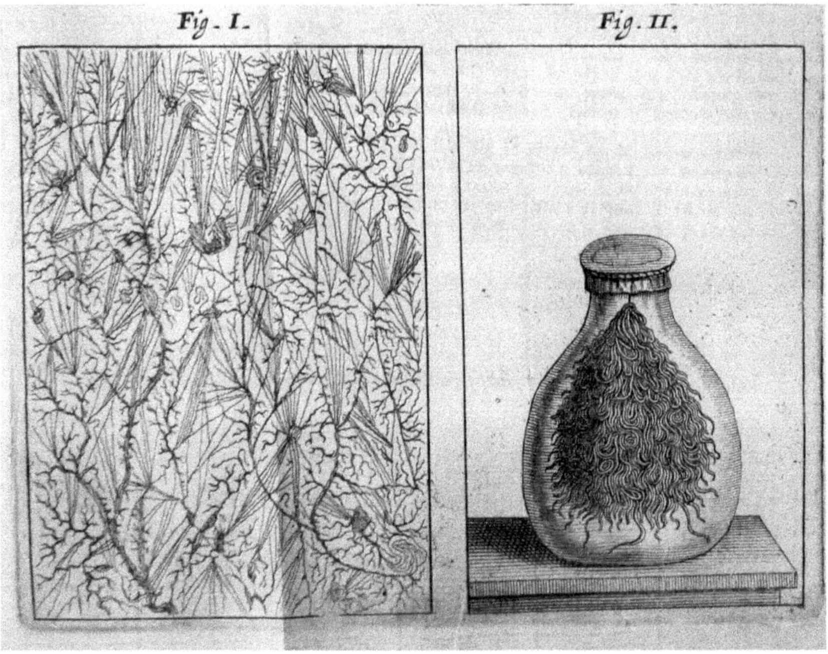

Figure 4.11: Anonymous, after Edmund King and Regnier de Graaf, *Plate for Issue 52*, Engraving, from *Philosophical Transactions of the Royal Society*, vol. 4, 1669. Natural History Museum Library, London via the Biodiversity Heritage Library.

list of suggested topics.[99] Beginning with the first letter Leeuwenhoek sent directly to Oldenburg, he followed Oldenburg's advice and included figures, in this case related to his discussion of the compression of air.[100]

Leeuwenhoek sent his great breakthrough with respect to generation just months after Oldenburg's death in September 1677.[101] In a paper he had translated into Latin, Leeuwenhoek detailed both his research methods and his amazing finding with regard to his examination of semen.[102] Unlike the previous examples where advances in understandings of the testicles were

99 *Correspondence of Henry Oldenburg*, vol. 9, 654.

100 Antoni van Leeuwenhoek to Henry Oldenburg, 15 August 1673, Archives of the Royal Society, Early Letters, EL/L1/1. Oldenburg translated this letter into English and published it in the *Philosophical Transactions*. "Considerations of Mr. Leewenhoeck," 21–23. For a detailed discussion of Leeuwenhoek's working methods and the importance of drawing to his practice of observing, see: Fransen, "Antoni van Leeuwenhoek," 485–544.

101 Leeuwenhoek sent the letter, which is not extant, to William Brouncker, who was President of the Royal Society at the time. Antoni van Leeuwenhoek, *Alle de brieven*, vol. 2, 276–299.

102 For the letter to Brouncker that possibly was a cover for the letter describing the discovery of sperm see: *Alle de brieven*, vol. 2, 272–275.

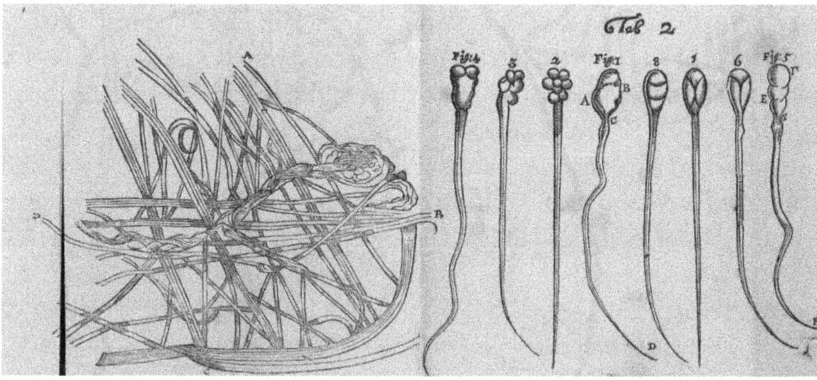

Figure 4.12: Anonymous, after Antoni van Leeuwenhoek, *Illustration for Issue 142*, Woodcut, from *Philosophical Transactions of the Royal Society*, vol. 12, 1678. © The Royal Society.

published first in books or broadsides and then debated and refined within the pages of the *Philosophical Transactions*, in this case, Leeuwenhoek's discovery was published first in the journal.[103] (Fig. 4.12) While Leeuwenhoek may not have been particularly surprised by his discoveries, others certainly were, and anatomical and microscopical research on generation continued.[104]

Over the course of Oldenburg's tenure as editor of the *Philosophical Transactions*, the journal's role in scholarly debate shifted. Where the first articles regarding mammalian reproduction presented research conducted in the decades before the founding of the Royal Society, Antoni van Leeuwenhoek's discovery of spermatozoa was published first in the *Philosophical Transactions*. Throughout these debates the circulation of visual evidence was essential to the arguments the authors were making. What they saw and how they managed to see it were key topics. Throughout these debates the engraver's burin was a key instrument as it allowed scholars separated by great distances to see what their interlocutors saw.

Conclusion

In the preface to the first issue of the second volume of the *Philosophical Transactions*, Oldenburg begged leave to "reflect a little upon what hath past."[105] He felt safe in assuming "that in these *Fragments*, something hath

103 Leeuwenhoek, "Observationes," 1040–1046.
104 Ruestow, *Microscope in the Dutch Republic*, 217.
105 Oldenburg, "Preface to the Third Year," 409.

been contributed to sowe such seeds, as may somewhat conduce to the illustration and improvement of Philosophy, and of all Laudable and Useful Arts and Practices."[106] As these two case studies of "the present figure of Saturn" and "the Parts of Generation" have shown, the *Philosophical Transactions* provided a pan-European community with a forum for recording and debating observations and experiments. For Oldenburg and his readers, each of "these Fragments" allowed for the slow accrual of information that eventually yielded accurate illustrations of Saturn and testicles, macroscopic and microscopic materials, and in so doing, revealed the significance of pursuing a consistent program of excellence—indeed, of accuracy—across fields that would in the future diverge widely. To press Oldenburg's metaphor, the pages of the *Philosophical Transactions* proved to be a fertile medium for the seeds sent by Oldenburg's correspondents. Despite some false starts after Oldenburg's death, the pages of the *Philosophical Transactions* continued, and indeed continue still, to be a repository for the accumulation of reports of experimental evidence, the refinement of previous findings and the first publication of research breakthroughs.

Bibliography

I. Manuscript Sources

London, Archives of the Royal Society
CLP/8i. Classified Papers. Astronomy.
EL/B1. Early Letters. Ball Letters.
EL/H2. Early Letters. Letters of Johannes Hevelius.
EL/L1–4. Early Letters. Letters of Antoni van Leeuwenhoek.

II. Printed Primary Sources

"An Accompt of the Improvement of Optick Glasses." *Philosophical Transactions* 1, no. 1 (6 March 1665): 2–3.

"An Account Concerning a Woman having a Double Matrix; as the Publisher hath Englished it out of the French, lately printed at Paris, where the Body was opened." *Philosophical Transactions* 4, no. 48 (21 June 1669): 969–970.

"Account of Some Books." *Philosophical Transactions* 3, no. 34 (13 April 1668): 663–604[664].

106 Oldenburg, "Preface to the Third Year," 409.

Aubry, Claude [Vadlius Dathirius Bonglarus, pseud.]. "Testis Examinatus." *Philosophical Transactions* 3, no. 42 (14 December 1668): 843–844.

—. *Testis Examinatus*. Florence: Ex typographia Joannis Francisci Barbettis, 1658.

Ball, William. "Of an Observation, not long since made in England, of Saturn." *Philosophical Transactions*, 1, no. 9 (12 February1666): 152–153.

Birch, Thomas. *The History of the Royal Society of London*. London: Printed for A. Millar, 1756.

Cassini, Jean-Dominique. "An Extract of Signor Cassini's Letter concerning a Spot lately seen in the Sun; together with a remarkable Observation of Saturn, made by the same." *Philosophical Transactions* 11, no. 128 (25 September 1676): 689–690.

Clarke, Timothy. "A Letter, written to the Publisher by the Learned and Experienced Dr. Timothy Clarck, one of His Majesties Physitians in Ordinary, concerning some Anatomical Inventions and Observations, particularlly the Origin of the Injection into Veins, the Transfusion of Bloud, and the Parts of Generation." *Philosophical Transactions* 3, no. 35 (18 May 1668): [672]– 682.

"Directions for Observations and Experiments to be made by Masters of Ships, Pilots, and other fit Persons in their Sea-Voyages." *Philosophical Transactions* 2, no. 24 (8 April 1667): 433–42[4]8.

Durston, William. "A Narrative of a Monstrous Birth in Plymouth, Octob. 22. 1670; together with the Anatomical Observations, taken thereupon by William Durston Doctor in Physick, and communicated to Dr. Tim. Clerk." *Philosophical Transactions* 5, no. 65 (14 November 1670): 2096–2098.

Galilei, Galileo. *Le Opere di Galileo Galilei.* Edizione Nationale, edited by A. Favaro (Florence, 1890–1909, reprinted 1929–39 and 1964–66).

Hevelius, Johannes. "An Extract of a Letter of M. Hevelius, Written to the Publisher from Dantzick, August 17/27 1670. Concerning a New Star, Lately Discover'd in the Constellation of the Swan, Together with the Present Appearance of the Planet Saturn." *Philosophical Transactions* 5, no. 65 (14 November 1670): 2087–2091.

—. "An Extract of Two Letters of M. Hevelius, of June 19. and of Octob. 7. 1671; Containing Some of His Late Celestial Observations, Touching Saturn Obscur'd by the Moon; a Lunar Eclipse; and an Occultation of the First of the Satellits of Jupiter by the Shadow of this Planet; a Transit of Jupiter and the Moon; and a Late Appearance of Saturn: All Here Deliver'd in the Language, Wherein the Author Communicated It." *Philosophical Transactions* 6, no. 78 (18 December 1671): 3027–3033.

—. "Eclipsis Solis Anno 1675. die 23 Junii mane st. n. observ. Gedani à Joh. Hevelio." *Philosophical Transactions* 11, no. 127 (18 July 1676): 660–661.

Hooke, Robert. "An Appendix to the Directions for Seamen, bound for far Voyages." *Philosophical Transactions* 1, no. 9 (12 February 1666): 147–149.

—. “The Particulars. Of Those Observations of the Planet Mars, Formerly Intimated to Have Been Made at London in the Months of February and March A[nno]. 1665/6,” *Philosophical Transactions*, 1, no. 14 (2 July 1666): 239–242.
Huygens, Christiaan. *Œuvres Complétes de Christiaan Huygens*. The Hague: Martinus Nijhoff, 1893.
—. *Systema Saturnium, sive, De Causis Mirandorum Saturni Phænomenôn*. The Hague: Ex Typographia Adriani Vlacq, 1659.
Huygens, Christiaan and Jean Picard. “Observation de Saturne faite à la Bibliothèque du Roy.” *Journal des Sçavans* 5, (11 February 1669): 11–12.
—. “An Observation of Saturne, made at Paris, the 17th of August, 1668. at hor. 11 1/2, at night, by M. Hugens, and M. Picart; as ‘tis describ’d in the Journal des Scavans of Febr. 11.1669.” *Philosophical Transactions* 4, no. 45 (25 March 1669): 900–901.
King, Edmund and Regnier de Graaf. “Some Observations Concerning the Organs of Generation, made by Dr. Edmund King, Fellow of the R. Society, and by Dr. Regnerus de Graeff, Physitian in Holland, which later occasioned the publishing of the former.” *Philosophical Transactions* 4, no. 52 (17 October 1669): [1043]–1047.
Leeuwenhoek, Antoni van. *Alle De Brieven Van Antoni Van Leeuwenhoek*. Edited by Gérard van Rijnberk. Amsterdam: N.V. Swets & Zeitlinger, 1939.
—. “Considerations of Mr. Leewenhoeck, touching the Compression of the Air; sent to the Publisher in his Letter of August 15. 1673.” *Philosophical Transactions* 9, no. 102 (27 April 1674): 21–23.
—. “Observationes D. Anthonii Lewenhoeck, de Natis è semine genitali Animalculis.” *Philosophical Transactions* 12, no. 142 (Dec.-Feb., 1678/9): 1040–1046.
Oldenburg, Henry. *Correspondence of Henry Oldenburg*. Edited by A. Rupert Hall and Marie Boas Hall. Madison: University of Wisconsin Press, 1965, vol. 1–12.
—. *Correspondence of Henry Oldenburg*. Edited by. A Rupert Hall and Maria Boas Hall. London and Philadelphia: Taylor & Francis, 1986, vol. 13.
—. “The Introduction.” *Philosophical Transactions* 1, no. 1 (6 March 1665): 1–2.
—. “A Preface to the Third Year of these Tracts.” *Philosophical Transactions* 2, no. 23 (11 March 1666/7): 409–415.
—. “To the Royal Society.” *Philosophical Transactions* 1, no. 1 (6 March 1665): n.p.
“*Philosophical Transactions*. A Londres, chez Jean Martin & James Allistry, Imprimeurs de la Societé Royale, & se trouve à Paris chez Jean Cusson, rue S. Jacques.” *Journal des Sçavans* 1 (30 March 1665): 156.
Placciu, Vincent. *De Scriptis & Scriptoribus Anonymis Atque Pseudonymis*. Hamburg: Sumptibus Christiani Guthii, 1674.
“Ragguaglio de Nuove Osservationi, da Giuseppe Campani. in. 12. In Roma.” *Journal des Sçavans* 1 (5 January 1665): 3–5.

III. Secondary Sources

Adams, J.C. "Note on William Ball's Observations of Saturn." *Monthly Notices of the Royal Astronomical Society*, 43 (1883): 92–97.

Andrade, E. N. da C. "The Birth and Early Days of the Philosophical Transactions." *Notes and Records of the Royal Society of London* 20, no. 1 (1965): 9–27.

Armitage, Angus. "William Ball. F.R.S. (1627–1690)." *Notes and Records of the Royal Society of London* 15 (July 1960): 167–172.

Atkinson, Dwight. "The *Philosophical Transactions of the Royal Society of London*, 1675–1975: A Sociohistorical Discourse Analysis." *Language in Society* 25 (1996): 333–371.

Bazerman, Charles. *Shaping Written Knowledge: The Genre and Activity of the Experimental Article*. Madison, WI: 1988.

Broman, Thomas. "Criticism and the Circulation of News: The Scholarly Press in the late Seventeenth Century." *History of Science* 51 (2013): 1–26.

Brown, Harcourt. *Scientific Organizations in Seventeenth Century France*. New York: Russell & Russell, 1967.

—. "History and the Learned Journal." *Journal of the History of Ideas* 33, no. 3 (1972): 365–378.

Conforti, Maria. "The Experimenters' Anatomy." In *The Accademia del Cimento and its European Context*. Edited by Marco Beretta, Antonio Clericuzio, and Lawrence M. Principe, 31–44. Sagamore Beach: Science History Publications/USA: 2009.

Doherty, Meghan C. "Giving Light to Narrative: The Use of Images in Early Modern Journals." *Nuncius: Journal of the Material and Visual History of Science* 30, no. 3 (2015): 543–569.

—. "'Ordinary skill in Cutts:' Visual Translation in Early Modern Learned Journals," in *Translation and the Circulation of Knowledge in Early Modern Science*, edited by Niall Hodson, Sietske Fransen, and Karl Enenkel. Leiden: Brill, 2017.

Fransen, Sietske. "Antoni van Leeuwenhoek, His images and Draughtsmen." *Perspectives on Science* 27, no. 3 (May–June 2019): 485–544.

Goodwin, Gordon. "Clarke, Timothy (*d.* 1672)," rev. Michael Bevan, *Oxford Dictionary of National Biography*, Oxford University Press, 2004; online ed., Jan 2008.

Gross, Joseph. "Ball, William (c.1631–1690)." *Oxford Dictionary of National Biography* (Oxford University Press, 2004; online ed. May 2006).

Hall, A. Rupert. "Medicine and the Royal Society." In *Medicine in Seventeenth Century England* edited by Allen G. DeBus. Berkeley: University of California Press, 1974, 421–452.

Hunter, Matthew C. *Wicked Intelligence: Visual Art and the Science of Experiment in Restoration London*. Chicago: University of Chicago Press, 2013.

Johns, Adrian. *The Nature of the Book: Print and Knowledge in the Making*. Chicago: University of Chicago Press, 1998.

—. "Miscellaneous Methods: Authors, Societies and Journals in Early Modern England." *The British Journal for the History of Science* 33, no. 2 (2000): 159–86.

Kronick, David A. *A History of Scientific & Technical Periodicals: The Origins and Development of the Scientific and Technical Press, 1665–1790*, 2nd ed. Metuchen, N.J.: 1976.

—. "Notes on the Printing History of the Early *Philosophical Transactions*." *Libraries and Culture* 25, no. 2 (1990): 243–268.

Lloyd, G.E.R. "Saving the Appearances." *The Classical Quarterly* 28, no. 1 (1978): 202–222.

McCutcheon, Roger Philip. "The *Journal Des Scavans* and the *Philosophical Transactions of the Royal Society*." *Studies in Philology* 21, no. 4 (1924): 626–628.

Moxham, Noah. "Fit for Print: Developing an Institutional Model of Scientific Periodical Publishing in England, 1665–ca.1714." *Notes and Records of the Royal Society* 69 (2015): 241–260.

—. "Authors, Editors and Newsmongers: Form and Genre in the *Philosophical Transactions* under Henry Oldenburg." In *News Networks in Early Modern Europe*, edited by Noah Moxham and Joad Raymond, 465–494. Leiden: Brill, 2016.

Ruestow, Edward G. *The Microscope in the Dutch Republic: The Shaping of Discovery*. Cambridge: Cambridge University Press, 1996.

Shapin, Steven, and Simon Schaffer. *Leviathan and the Air-Pump: Hobbes, Boyle, and the Experimental Life*. Princeton, N.J.: Princeton University Press, 1985.

Van Helden, Albert. "'Annulo Cingitur': The Solution of the Problem of Saturn." *Journal for the History of Astronomy*, 5 (1974): 155–74.

—. "Christopher Wren's '*De Corpore Saturni*'." *Notes and Records of the Royal Society of London*, 23, no. 2 (Dec. 1968): 213–229.

—. "Saturn and his Anses." *Journal of the History of Astronomy* 5 (1974): 105–121.

Winkler, Mary G., and Albert van Helden. "Representing the Heavens: Galileo and Visual Astronomy." *Isis* 83, no. 2 (1992): 195–217.

—. "Johannes Hevelius and the Visual Language of Astronomy." In *Renaissance and Revolution: Humanists, Scholars, Craftsmen, and Natural Philosophers in Early Modern Europe*, edited by Judith Veronica Field and Frank A. J. L. James, 95–114. Cambridge: Cambridge University Press, 1993.

Conclusion

The four chapters of this book have traced a complex web of interrelationships among people, material objects, drawings, and intaglio printed images, all of which were actively involved in the development of understandings of accuracy in the period. As these chapters have shown, these relationships were not simply binaries. Prints and drawings were not related through a logic of exclusion, that is, the carving of information into a copper plate did not negate the value of the drawing on which the print was based. Instead, the durable interdependencies of different modes of representation put pressure on an understanding of the relationship between prints and drawings as adversarial. As the chapters on Willughby's *Ornithology* and the *Philosophical Transactions* show, drawings were not necessarily the preliminary step toward creating a printed image. Drawings were made after prints and prints were made after other prints. Prints and drawing were intricately wound together in the early visual culture of the Royal Society. Further, I have stressed over these four chapters that their relationship was neither adversarial nor linear. Both drawings and prints were essential to the visual communication of knowledge by Fellows of the Royal Society and their correspondents, and the same care that went into performing experiments also went into creating images.

Following the work of Deleuze and Guattari, I have treated each of the works considered in this study as rhizomatic structures that exceed their bindings; that is, as interconnected networks of information.[1] The works considered are intertextual and intermedial—they are interconnected with other texts, images, and objects and, thus, the chapters were necessarily "comparative" in that they set the books in a dense web of other books, single-sheet prints, and drawings. The chapters traced the complex interrelationships between material objects and images by both looking closely at the resulting printed images and by systematically tracing the source material for those images. The intermediality of the works under consideration was not proposed in abstract terms, but rather in concrete

1 Deleuze and Guattari, "Rhizome," in *On the Line*, 22.

Doherty, M.C., *Engraving Accuracy in Early Modern England: Visual Communication at the Early Royal Society*. Taylor & Francis Group, 2022.
DOI 10.5117/9789463721066_CONCL

historicized terms allowing for a deeper understanding of how accuracy was produced at a given moment in the past.

By including the burin in the pantheon of scientific instruments, I have thus argued that illustration was integral to the experiments and their communication. The visual effects of accuracy, then, were not dependent on a linear relationship between circulating prints and source drawings, in which only a print that visually signified its dependence on a drawing could be accepted as accurate. Instead, as discussed throughout this book, a set of protocols existed that produced the effect of accuracy. Chapter one discussed protocols that disciplined the hand and the eye of the young gentleman as he learned to draw and engrave. Text and images worked together to develop careful work habits. Visual judgment developed through a combination of historical and practical knowledge. The writings of William Faithorne and Robert Boyle, examined in chapter one, exemplify further disciplinary principles and highlight how printed books worked to control the actions of artisans and experimenters in order to produce outcomes with visual traces of accuracy. William Faithorne's writings also highlight the recursive aspect of this protocol as the same tool used to achieve the effect of accuracy, the burin, was used to teach accuracy. Chapters two and three investigated protocols in which the effect of accuracy was achieved by means of visual references to previous images, although quite different sources served as referents in these two chapters. Taken together chapters two and three highlight the multiplicity of sources that could be used to produce the effect of accuracy. The final chapter explored the role of consensus in the creation of an effect of accuracy. The circulation of images among Henry Oldenburg's correspondents and readers produced images that had the visual traces of accuracy making them useful to the Royal Society's research agenda. The four chapters of this book uncover different regimes of accuracy that complicate our understanding of what it meant for an image to be accurate by stressing that there was no single path to creating an authoritative image.

The regimes of accuracy traced in this study provided the groundwork for images to successfully transmit knowledge both within the close community of the Royal Society and more broadly through the pan-European scholarly community. With the decline of the use of Latin for scholarly communication the visual became an open form of communication that overcame linguistic barriers. As discussed in chapter four, then Antoni van Leeuwenhoek began corresponding with Henry Oldenburg, Oldenburg specifically asked for images to be included with his next letter.[2] Oldenburg then had the drawings

2 *Correspondence of Henry Oldenburg*, vol. 9, 654.

Leeuwenhoek sent engraved to accompany his text in the *Philosophical Transactions*. The engraved line became a new *lingua franca* for scientific communication as images replaced Latin as the predominate vernacular for open communication.

Printed images that exhibited the visual traces of accuracy were dependent on the careful actions of artisans and Fellows of the Royal Society. Fellows acknowledged this need for care and were interested in the production of both experiments and images. As discussed, John Evelyn dedicated his history of engraving, *Sculptura: Or the History, and Art of Chalcography and Engraving in Copper,* to "The Honourable, and Learned Gentleman, Robert Boyle Esq." Evelyn and Boyle were both founding members of the Royal Society, and both were involved in projects relating to the arts and to natural philosophy. Evelyn began his dedicatory epistle by acknowledging Boyle's role in his pursuing the history of engraving: "Having upon your reiterated instances (which are ever commands with me) prepared this Treatise concerning the History of *Chalcography &c.*"(A2r) He then went on to recount some of the reasons why Boyle encouraged him to publish this book:

> ...but as you are pleased to judge it useful for the encouragement of the Gentlemen of our Nation, who sometimes please themselves with these innocent diversions (Collections worthy of them for divers respects) and, especially, that such as are addicted to the more Noble *Mathematical Sciences*, may draw, and engrave their *Schemes* with delight and assurance, I have been induc'd to think it more worthy your Patronage, and of my small Adventure. (A2v)

Evelyn saw his "small Adventure" as being a worthy endeavor because it benefited both collectors of prints, because it provided a history of engraving, and those "addicted to the more Noble *Mathematical Sciences*." Evelyn saw his book as meeting a demand from within his community of Fellows. In particular, Evelyn established his audience as those engaged in what we would now classify as the physical sciences. His text, he asserted, would confidently allow astronomers and mathematicians to create their own images. Although *Sculptura* does not explicate a detailed method, like *A Book of Drawing* or Faithorne's *The Art of Graveing*, Evelyn views his book as providing natural philosophers with the information necessary to draw and engrave the plates for their books. Furthermore, Evelyn saw his book as: "...contributing to that great and august designe, which your [Boyle's] illustrious, and happy *Genius* do's prompt you to, of cultivating the *Sciences*, and advancing of useful knowledge." (A2v-A3r) *Sculptura* then was not only beneficial to those with an interest in collecting prints, but to those who, along with Boyle and Evelyn, worked for the advancement of learning as encouraged by the writings of Francis Bacon. In this dedication, Evelyn tied learning to draw and engrave directly to the larger goals of the Royal Society and clearly outlined the shared habitus between artisans and experimenters. Working knowledge of drawing and engraving was understood as components of the overarching goal of the Royal Society to promote experimental learning.

Connecting drawing and engraving to the creation and circulation of knowledge in the context of the early Royal Society is not an anachronistic gesture. These four chapters work to re-establish the connection between art and science that was critical to the "promotion of experimental learning" for Fellows of the Royal Society. The efforts of artisans and Fellows to establish methods for the production of reproducible results were not discrete endeavors pursued by unconnected groups of people. Instead the production of printed images that had the traces of careful actions was the result of attention to accuracy among these groups. That is, engravers worked to produce images that carefully translated into print the observations and experimental results of natural historians and natural philosophers. For their part, Fellows of the Royal Society sought to publish books that faithfully recorded their research. In the case of Willughby's *Ornithology*, John Ray combed through previously printed books, a huge collection of drawings, Willughby's notes, as well as his own, and corresponded with a wide array of people in order to provide the engravers with a group of images which served as the basis for the printed images included in his book, and it was his efforts as well as the engravers' that made it possible for him to describe the images as "nearly resembling the live birds." All three of the case study chapters in this book explore how the actions of artisans and Fellows produced images that had the visual traces of accuracy, and the first chapter established how methods were developed to train the viewer to look for and see through the "innocent *Witch-craft* of lights and shades."

This shared attention to accuracy leads me to claim the engraver's burin, or graver, should be considered among the pantheon of instruments that fundamentally changed how nature was studied and subsequently understood in the seventeenth century. The burin should be considered alongside the microscope and the air-pump as key instruments for the advancement of learning. In making this assertion, I am building on the work of Thomas Hankins and Robert Silverman who argue that one of the key features of philosophical instruments in the seventeenth century was that they distorted the object of study. In discussing the microscope and the air-pump, they state:

> Instead of just measuring length, weight, or time, these instruments distorted nature in some way, either by magnifying it as in the case of the telescope and microscope, or by producing an unnatural condition as in the vacuum created in an air pump.[3]

Like these well-known seventeenth-century instruments, the engraver's burin distorted nature by making it two-dimensional and black and white. There were a series of aesthetic decisions that went into the creation of an image that distorted nature to ease the communication process. Color and texture were suggested with a combination of lines. Much like how methods for the use of the air-pump and microscope were written about by their main advocates within the Royal Society, Robert Boyle on the air-pump and Robert Hooke on the microscope, the method for using the burin was written about by a skilled user, William Faithorne. With all three of these instruments, users tried to manage the distortion created by their use through careful actions. My goal in arguing that the burin be considered alongside the microscope and the air-pump is twofold: first, I want to stress that accuracy in printed images in the seventeenth century was produced through a series of distortions and decisions, and the effect of accuracy was the production of authoritative images. Second, I want to collapse further the perceived distance between art and science in the early modern period. This perceived distance between art and science is a modern construction, which is projected onto historical periods where there was no such distance.

Throughout this book I have consciously and consistently brought together seemingly disparate groups of people, images, and objects to expand our understandings of how images are made and why they are important to study. The result is a textured portrait of a series of complex interrelationships that underscores the importance of thinking through and with networks to unearth the care with which images were created in the context of the early Royal Society of London. While the Royal Society has served as the locus for this study, the arguments made and the discoveries explicated in this project have implications beyond the specific context of London in the first years after the Restoration. Attention to careful actions continues to be essential to the communication of knowledge. Although printed images

3 Hankins and Silverman, *Instruments and the Imagination*, 3.

may be on the verge of extinction in scientific communication, accuracy remains a concern among scientists who continue to use images in their research and publications.

Bibliography

Evelyn, John. *Sculptura: Or the History, and Art of Chalcography and Engraving in Copper.* London: Printed by J. C. for G. Beedle & T. Collins, 1662.

Deleuze, Gilles, and Felix Guattari. *On the Line.* Translated by John Johnston. New York: Semiotexte, 1983.

Hankins, Thomas L., and Robert J. Silverman. *Instruments and the Imagination.* Princeton, N.J.: Princeton University Press, 1995.

Oldenburg, Henry. *Correspondence of Henry Oldenburg.* Edited by A. Rupert Hall and Marie Boas Hall. Madison: University of Wisconsin Press, 1965, vol. 1–12.

Bibliography

I. Manuscript Sources

London, Archives of the Royal Society

CLP/2. Classified Papers. Surveying, Opticks, Perspective, Sculpture, Painting, Music, Acoustics Mechanicks.

CLP/8i. Classified Papers. Astronomy.

CLP/20. Classified Papers. Hooke Papers.

CLP/24. Classified Papers. Collins, Oldenburg, Hooke.

EL/B1. Early Letters. Ball Letters.

EL/E. Early Letters. Letters of John Evelyn.

EL/H2. Early Letters. Letters of Johannes Hevelius.

EL/L1–4. Early Letters. Letters of Antoni van Leeuwenhoek.

EL/W3. Early Letters. Letter of Francis Willughby.

LBO/1. Letter Book Original. 1661–1666.

LBO/2. Letter Book Original. 1667–1668.

MS215. Copies of Papers read to the Society about 1662–1664.

RBO/2i. Register Book Original.

London, British Library

Add. MS 5260. Drawings of Animals & some parts of Men for comparative Anatomy by Dr.

Sam.l Collins & Dr Edward Tyson. Bought of Dr Collins's Daughter.

Add. MS 5263. Drawings of Birds in Colors, by G. Edwards, and others, being the originals of his work on the Natural History of Birds. 4pls. 4to. Lond. 1743–50. (154 in number.)

Add. MS 5264. A similar Volume (144 in number.)

Add. MS 5265. A similar Volume (136 in number.)

Add. MS 5266. Drawings of Birds in Colors, crayons, and pencil, by Dr. Browne, and others (163 in number.)

Add. MS 6485. An exact natural Description and Deliniation of the Water-Fowls, Fishes,

Fourfooted Beasts, Insects, and Worms, which are found in the water at Strassburg, which I myself shot and handled. Leonhard Baldner, Fisher and Citizen in Strassburg.

Add. MS 57495. John Covel, D.D., (b.1638, d.1722): natural history notebook and commonplace book; 1660–1713, n.d. Contains many pen and ink drawings of plants, insects, fish, etc. with detailed descriptive text.

Add. MS 78340. Evelyn Papers. Vol. CLXXIII. Collection in two parts, the first of medical and culinary recipes, the second of skills and crafts; 1659–[early 18th cent?].

Nottingham, University of Nottingham

Mi LM 24. Collection of water-colour drawings and engravings of birds. N.d. mid-17th-early 18th century.

Providence, R.I., Brown University

QL41 .B3 1653 v.1–4. Recht natÿrliche Beschreibung und Abmalung der Wasser Vögel, Fischen, Vierfüssigen Thieren, Insecten, und Gewürm, so beÿ Strassburg in den Wassern

gefunden werden, welche ich selber geschossen, gefangen und in mein eigner Hand gehabt.

II. Printed Primary Sources

Academia Italica, The Publick School of Drawing of the Gentlemans Accomplishment. London: Printed by P. Lillicrap, 1666.

"An Accompt of the Improvement of Optick Glasses." *Philosophical Transactions* 1, no. 1 (6 March 1665): 2–3.

"An Account Concerning a Woman having a Double Matrix; as the Publisher hath Englished it out of the French, lately printed at Paris, where the Body was opened." *Philosophical Transactions* 4, no. 48 (21 June 1669): 969–970.

"Account of Some Books." *Philosophical Transactions* 3, no. 34 (13 April 1668): 663–604[664].

"An Account of Some Books." *Philosophical Transactions* 3, no. 39 (21 September 1668): 779–88.

Aldrovandi, Ulisse. *Ornithologiae, Hoc est De Avibus Historiae Libri XII.* Bologna: Apud Franciscum de Franciscis Senensem, 1599–1603.

Aubrey, John. *"Brief Lives," Chiefly of Contemporaries, Set Down by John Aubrey, between the Years 1669 and 1696.* Edited by Andrew Clark. Oxford: Clarendon Press, 1898.

Aubry, Claude [Vadlius Dathirius Bonglarus, pseud.]. *Testis Examinatus.* Florence: Ex typographia Joannis Francisci Barbettis, 1658.

—. "Testis Examinatus." *Philosophical Transactions* 3, no. 42 (14 December 1668): 843–44.

Ball, William. "Of an Observation, not long since made in England, of Saturn." *Philosophical Transactions*, 1, no. 9 (12 February 1666): 152–153.

Bate, John. *The Mysteryes of Nature and Art: Conteined in Foure Severall Tretises, the First of Water Workes, the Second of Fyre workes, the Third of Drawing, Colouring, Painting, and Engraving, the Fourth of Divers Experiments, as wel Serviceable as Delightful.* 2nd edition. London: Printed for Ralph Mab, 1635.

Birch, Thomas. *The History of the Royal Society of London.* London: Printed for A. Millar, 1756.

Bondt, Jakob de. "Historiæ Naturalis and Medicæ Indiæ Orientalis Libri Sex." In *De Indiœ Utriusque Re Naturali et Medica Libri Quatuordecim*, edited by Willem Piso. Amsterdam: Apud Ludovicum et Danielem Elzevirios, 1658.

A Book of Drawing, Limning, Washing or Colouring of Maps and Prints, or, the Young-Man's Time Well Spent. London: Printed by James and Joseph Moxon, for Thomas Jenner, 1647.

A Book of Drawing, Limning, Washing or Colouring of Maps and Prints: And the Art of Painting, with the Names and Mixtures of Colours used by the Picture-Drawers. Or, The Young-mans Time well Spent. London: Printed by M. Simmons, for Thomas Jenner, 1666.

A Booke of Drawinges. Performed according to the best order for use & Brevity that is yet Extant. London: Printed and are to be sould by Peter Stent, 1650.

A Booke of Portraicture. London: Sold by Godfrey Richards, 1665.

Borel, Pierre. *Observationum Microcospicarum Centuria.* The Hague: Ex typographia Adriani Vlacq, 1655.

Bosse, Abraham. *Moyen Universel de Practiquer la Perspective sur les Tableaux, ou Surfaces Irregulieres.* Paris: Chez ledit Bosse, 1653.

—. *Traicte des Manieres de Graver en Taille Douce sur l'Airin.* Paris: Chez Bosse, 1645.

Boyle, Robert. *Certain Physiological Essays, Written at Distant Times, and on Several Occasions.* London: Printed for Henry Herringman, 1661.

—. *A Continuation of New Experiments, Physico-Mechanical, Touching the Spring and Weight of the Air, and Their Effects.* Oxford: Printed by Henry Hall printer to the University, for Richard Davis, 1669.

—. *The Correspondence of Robert Boyle.* Edited by Michael Cyril William Hunter, Antonio Clericuzio and Lawrence Principe. 6 vols. London: Pickering & Chatto, 2001.

—. *Medicina Hydrostatica, or, Hydrostaticks Applyed to the Materia Medica.* London: Printed for Samuel Smith, 1690.

—. *New Experiments Physico-Mechanicall, Touching the Spring of the Air, and Its Effects, Made, for the Most Part, in a New Pneumatical Engine.* Oxford: Printed by H: Hall for Tho: Robinson, 1660.

—. *Some Considerations Touching the Usefulnesse of Experimental Naturall Philosophy.* Oxford: Printed by Henry Hall, 1664.

Brown, John. "A Remarkable Account of a Liver, Appearing Glandulous to the Eye." *Philosophical Transactions of the Royal Society* 15, no. 178 (1685): 1266–1268.

Browne, Alexander. *Ars Pictoria: Or an Academy Treating of Drawing, Painting, Limning, and Etching*. London: Printed by J. Remayne for the Author, 1669.

—. *A Compendious Drawing-Book*. London: Printed for Austin Oldisworth, 1677.

—. *The Whole Art of Drawing, Painting, Limning, and Etching*. London: Printed for Peter Stint, 1660.

Buonanni, Filippo. *Micrographia Curiosa sive Rerum Minutissimarum Observations*. Rome: Typis Dominici Antonij Herculis, 1691.

Casseri, Gulio. *Pentaestheseion, hoc est De qvinqve sensibvs liber*. Venice: Apud Nicolaum Misserinum, 1609.

Cassini, Jean-Dominique. "An Extract of Signor Cassini's Letter concerning a Spot lately seen in the Sun; together with a remarkable Observation of Saturn, made by the same." *Philosophical Transactions* 11, no. 128 (25 September 1676): 689–690.

Clarke, Timothy. "A Letter, written to the Publisher by the Learned and Experienced Dr. Timothy Clarck, one of His Majesties Physitians in Ordinary, concerning some Anatomical Inventions and Observations, particularlly the Origin of the Injection into Veins, the Transfusion of Bloud, and the Parts of Generation." *Philosophical Transactions* 3, no. 35 (18 May 1668): [672]–682.

Clusius, Carolus. *Exoticorum Libri Decem*. Antwerp: Ex Officinâ Plantinianâ Raphelengii, 1605.

Collins, Samuel. *A Systeme of Anatomy, Treating of the Body of Man, Beasts, Birds, Fish, Insects, and Plants*. London: Printed by Thomas Newcomb, 1685.

Cousin, Jean. *Livre de Pourtraittre de Maistre Iean Cousin Peintre et Geometrien Tres-Excellent*. Paris: Chez Jean le Clerc, 1595.

Derham, William. *Select Remains of the Learned John Ray: With His Life*. London: Printed and sold by J. Dodsley and J. Walter, 1760.

Diagraphia, sive Ars Delineatoria. Amsterdam: Joannes Janssonius, 1616.

"Directions for Observations and Experiments to be made by Masters of Ships, Pilots, and other fit Persons in their Sea-Voyages." *Philosophical Transactions* 2, no. 24 (8 April 1667): 433–42[4]8.

Dürer, Albrecht. *Underweysung der messung mit dem zirckel un richt scheyt*. Nuremberg: I. Formschneyder, 1525.

Durston, William. "A Narrative of a Monstrous Birth in Plymouth, Octob. 22. 1670; together with the Anatomical Observations, taken thereupon by William Durston Doctor in Physick, and communicated to Dr. Tim. Clerk." *Philosophical Transactions* 5, no. 65 (14 November 1670): 2096–2098.

Edwards, George. *A Natural History of Birds*. London: Printed for the Author, 1743.

Evelyn, John. *The Diary of John Evelyn*. Edited by E.S. de Beer. Oxford: Clarendon Press, 1955.

—. *An Idea of the Perfection of Painting.* London: Printed for Henry Herringman, 1668.

—. "A Letter...Concerning the Spanish Sembrador or New Engin for Ploughing, and Equal Sowing All Sorts of Grain, and Harrowing, at Once." *Philosophical Transactions* 5, no. 60 (20 June 1670): 1055–65.

—. *A Parallel of the Antient Architecture with the Modern.* London: Printed by Tho. Roycroft for John Place, 1664.

—. *Sculptura: Or the History, and Art of Chalcography and Engraving in Copper.* London: Printed by J. C. for G. Beedle & T. Collins, 1662.

The Excellency of the Pen and Pencil, Exemplifying the Uses of them in the most Exquisite and Mysterious Art of Drawings, Etching, Engraving, Limning. Painting in Oyl, Washing of Maps and Pictures. London: Printed by Thomas Ratcliff and Thomas Daniel, for Dorman Newman and Richard Jones, 1668.

Faithorne, William. *The Art of Graveing, and Etching, Wherein Is Exprest the True Way of Graveing in Copper.* London: Published by the Author, 1662.

Galilei, Galileo. *Le Opere di Galileo Galilei.* Edizione Nationale, edited by A. Favaro (Florence, 1890–1909, reprinted 1929–39 and 1964–66).

Gessner, Konrad. *Historiæ Animalium Liber III.* Zurich: C. Froschouerum, 1555.

Grew, Nehemiah. *Musaeum Regalis Societatis, or, a Catalogue & Description of the Natural and Artificial Rarities Belonging to the Royal Society.* London: Printed by W. Rawlins, for the Author, 1681.

Griendel, Johann Franz. *Micrographia Nova, Sive, Nova et Curiosa Variorum Minutorum Corporum.* Nuremberg: Sumptibus Johannis Ziegeri, 1687.

Haydocke, Richard. *A Tracte Containing the Artes of Curious Paintinge, Carvinge & Buildinge.* Oxford: Joseph Barnes for R.H., 1598.

Hevelius, Johannes. "Eclipsis Solis Anno 1675. die 23 Junii mane st. n. observ. Gedani à Joh. Hevelio." *Philosophical Transactions* 11, no. 127 (18 July 1676): 660–661.

—. "An Extract of a Letter of M. Hevelius, Written to the Publisher from Dantzick, August 17/27 1670. Concerning a New Star, Lately Discover'd in the Constellation of the Swan, Together with the Present Appearance of the Planet Saturn." *Philosophical Transactions* 5, no. 65 (14 November 1670): 2087–2091.

—. "An Extract of Two Letters of M. Hevelius, of June 19. and of Octob. 7. 1671; Containing Some of His Late Celestial Observations, Touching Saturn Obscur'd by the Moon; a Lunar Eclipse; and an Occultation of the First of the Satellits of Jupiter by the Shadow of this Planet; a Transit of Jupiter and the Moon; and a Late Appearance of Saturn: All Here Deliver'd in the Language, Wherein the Author Communicated It." *Philosophical Transactions* 6, no. 78 (18 December 1671): 3027–3033.

—. *Selenographia; Sive, Lunae Descriptio.* Gdansk: Autoris sumtibus, Typis Hünefeldianis, 1647.

Hodierna, Giovanni Battista. *L'Occhio della Mosca Discorso Fisico*. Palermo: Per Decio Cirillo, 1644.

Hooke, Robert. "An Appendix to the Directions for Seamen, bound for far Voyages." *Philosophical Transactions* 1, no. 9 (12 February 1666): 147–149.

—. "A Contrivance to Make the Picture of Anything Appear on a Wall, Cup-board, or within a Picture-frame, &c. in the Midst of a Light Room in the Day-time; or in the Night-time in any Room that is Enlightned with a Considerable Number of Candles; Devised and Communicated by the Ingenious Mr. Hook." *Philosophical Transactions* 3, no. 38 (17 August 1668): 741–743.

—. *The Diary of Robert Hooke, M.A., M.D., F.R.S., 1672–1680, Transcribed from the Original in the Possession of the Corporation of the City of London (Guildhall Library)*. Edited by Henry W. Robinson and Walter Adams. London: Taylor & Francis, 1935.

—. *Microscopic Observations or, Dr. Hooke's Wonderful Discoveries by the Microscope, Illustrated by Thirty-Three Copper-Plates, Curiously Engraved: Whereby the Most Valuable Particulars in That Celebrated Author's Micrographia Are Brought Together.* London: Printed for Robert Wilkinson, 1780.

—. *Micrographia, or, Some Physiological Descriptions of Minute Bodies Made.* London: Printed by J. Martyn and J. Allestry, 1665.

—. *Micrographia Restaurata or, the Copper-Plates of Dr. Hooke's Wonderful Discoveries by the Microscope, Reprinted and Fully Explained: Whereby the Most Valuable Particulars in That Celebrated Author's Micrographia Are Brought Together in a Narrow Compass*. Edited by Henry Baker. London: Printed for and sold by John Bowles, 1745.

—. "The Particulars. Of Those Observations of the Planet Mars, Formerly Intimated to Have Been Made at London in the Months of February and March A[nno]. 1665/6," *Philosophical Transactions*, 1, no. 14 (2 July 1666): 239–242.

Huygens, Christiaan. *Œuvres Complétes de Christiaan Huygens*. The Hauge: Martinus Nijhoff, 1893.

—. *Systema Saturnium, sive, De Causis Mirandorum Saturni Phænomenôn*. The Hague: Ex Typographia Adriani Vlacq, 1659.

Huygens, Christiaan and Jean Picard. "Observation de Saturne faite a la Bibliotheque du Roy." *Journal des Sçavans* 5, (11 February 1669): 11–12.

—. "An Observation of Saturne, made at Paris, the 17th of August, 1668. at hor. 11 1/2, at night, by M. Hugens, and M. Picart; as 'tis describ'd in the Journal des Scavans of Febr. 11.1669." *Philosophical Transactions* 4, no. 45 (25 March 1669): 900–901.

King, Edmund and Regnier de Graaf. "Some Observations Concerning the Organs of Generation, made by Dr. Edmund King, Fellow of the R. Society, and by Dr. Regnerus de Graeff, Physitian in Holland, which later occasioned the publishing of the former." *Philosophical Transactions* 4, no. 52 (17 October 1669): [1043]–1047.

Leeuwenhoek, Antoni van. *Alle De Brieven Van Antoni Van Leeuwenhoek.* Edited by Gérard van Rijnberk. Amsterdam: N.V. Swets & Zeitlinger, 1939.

—. "Considerations of Mr. Leewenhoeck, touching the Compression of the Air; sent to the Publisher in his Letter of August 15. 1673." *Philosophical Transactions* 9, no. 102 (27 April 1674): 21–23.

—. "Observationes D. Anthonii Lewenhoeck, de Natis è semine genitali Animalculis." *Philosophical Transactions* 12, no. 142 (Dec.–Feb., 1678/9): 1040–1046.

"Micrographia, or Some Physiological descriptions of minute Bodes made by magnifying glasses, with observations and inquires thereupon, by R. Hooke fellow the Royal Society. In fol. London." *Journal des Sçavans,* 2 (20 December 1666): 491–501.

Nieremberg, Juan Eusebio. *Historia Naturae, Maxime Peregrinae, Libris XVI.* Antwerp: Ex officina Plantiniana Balthasaris Moreti, 1635.

Oldenburg, Henry. "An Account of Some Books." *Philosophical Transactions* 10, no. 120 (27 December 1675): 481–90.

—. *Correspondence of Henry Oldenburg.* Edited by A. Rupert Hall and Marie Boas Hall. Madison: University of Wisconsin Press, 1965, vol. 1–12.

—. *Correspondence of Henry Oldenburg.* Edited by. A Rupert Hall and Maria Boas Hall. London and Philadelphia: Taylor & Francis, 1986, vol. 13.

—. "The Introduction." *Philosophical Transactions* 1, no. 1 (6 March 1665): 1–2.

—. "A Preface to the Third Year of these Tracts." *Philosophical Transactions* 2, no. 23 (11 March 1666/7): 409–415.

—. "To the Royal Society." *Philosophical Transactions* 1, no. 1 (6 March 1665): n.p.

Olina, Giovanni Pietro. *Uccelliera: Overo, Discorso della Natura e Proprieta di Diversi Uccelli.* Rome: A. Fei, 1622.

Peacham, Henry. *The Art of Drawing with the Pen, and Limning in Watercolours.* London: Richard Braddock for William Jones, 1607.

Pepys, Samuel. *The Diary of Samuel Pepys: A New and Complete Transcription.* Edited by Robert Latham and William Matthews. Berkeley: University of California Press, 1970.

"*Philosophical Transactions.* A Londres, chez Jean Martin & James Allistry, Imprimeurs de la Societé Royale, & se trouve à Paris chez Jean Cusson, rue S. Jacques." *Journal des Sçavans* 1 (30 March 1665): 156.

Piso, Willem. *Historia Naturalis Brasiliae.* Leiden: Apud Franciscum Hackium, 1648.

—. *De Indiæ Utriusque re Naturali et Medica Libri Quatuordecim.* Amsterdam: Apud Ludovicum et Danielem Elzevirios, 1658.

Placciu, Vincent. *De Scriptis & Scriptoribus Anonymis Atque Pseudonymis.* Hamburg: Sumptibus Christiani Guthii, 1674.

Power, Henry. *Experimental Philosophy, in Three Books: Containing New Experiments Microscopical, Mercurial, Magnetical.* London: Printed by T. Roycroft for John Martin and James Allestry, 1664.

"Ragguaglio de Nuove Osservationi, da Giuseppe Campani. in. 12. In Roma." *Journal des Sçavans* 1 (5 January 1665): 3–5.

Ray, John. *The Correspondence of John Ray*. Edited by Edwin Lankester. London: Printed for the Ray Society, 1848.

—. *Further Correspondence of John Ray*. Edited by Robert W. T. Gunther. London: Printed for the Ray Society, 1928.

—. *Observations Topographical, Moral, & Physiological Made in a Journey through Part of the Low-Countries, Germany, Italy, and France*. London: Printed for John Martyn, 1673.

Sprat, Thomas. *The History of the Royal-Society of London for the Improving of Natural Knowledge*. London: Printed by T. R. for J. Martyn and J. Allstrey, 1667.

Stelluti, Francesco. *Persio: Tradotto in Verso Sciolto e Dichiarato*. Rome: G. Mascardi, 1630.

Tyson, Edward. "Vipera Caudi-Sona Americana, or the Anatomy of a Rattle-Snake, Dissected at the Repository of the Royal Society in January 1682/3." *Philosophical Transactions of the Royal Society* 13, no. 144 (1683): 25–58.

Vesalius, Andreas. *De Humani Corporis Fabrica libri septem*. Basel: Ex officina Johannis Oporini, 1543.

Willughby, Francis, and John Ray. *Ornithologiae libri tres* London: Impensis Joannis Martyn, 1676.

—. *The Ornithology of Francis Willughby*. London: Printed by A.C. for John Martyn, 1678.

III. Secondary Sources

Adams, J.C. "Note on William Ball's Observations of Saturn." *Monthly Notices of the Royal Astronomical Society*, 43 (1883): 92–97.

Alexander, David. "Faithorne, Loggan Vandrebanc and White: The Engraved Portrait in Late Seventeenth-Century Britain." In *Printed Images in Early Modern Britain: Essay in Interpretation* edited by Michael Hunter, 297–316. Farham, Surrey: Ashgate, 2010.

Alpers, Svetlana. *The Art of Describing: Dutch Art in the Seventeenth Century*. Chicago: University of Chicago Press, 1983.

Andrade, E. N. da C. "The Birth and Early Days of the Philosophical Transactions." *Notes and Records of the Royal Society of London* 20, no. 1 (1965): 9–27.

Armitage, Angus. "William Ball. F.R.S. (1627–1690)." *Notes and Records of the Royal Society of London* 15 (July 1960): 167–172.

Atkinson, Dwight. "The *Philosophical Transactions of the Royal Society of London*, 1675–1975: A Sociohistorical Discourse Analysis." *Language in Society* 25 (1996): 333–371.

Baigrie, Brian S., ed. *Picturing Knowledge: Historical and Philosophical Problems Concerning the Use of Art in Science*. Toronto: University of Toronto Press, 1996.

Bazerman, Charles. *Shaping Written Knowledge: The Genre and Activity of the Experimental Article*. Madison, WI: 1988.

Bartrum, Giulia. *Albrecht Dürer and His Legacy: The Graphic Work of a Renaissance Artist*. Princeton, N.J.: Princeton University Press, 2002.

Barzman, Karen-edis. *The Florentine Academy and the early modern state: The discipline of disegno*. Cambridge: Cambridge University Press, 2000.

Bell, C.F. and Rachel Poole. "English Seventeenth-Century Portrait Drawings in Oxford Collections: Part II Miniature Portraiture in Black Lead and Indian Ink." *Walpole Society* 14 (1925–26): 43–80.

Bennett, Jim. "Hooke's Instruments," in *London's Leonardo: The Life and Work of Robert Hooke*, 63–104. Oxford: Oxford University Press, 2003.

Bermingham, Ann. *Learning to Draw: Studies in the Cultural History of a Polite and Useful Art*. New Haven: Yale University Press, 2000.

Birkhead, Tim, ed. *Virtuoso by Nature: The scientific worlds of Francis Willughby FRS (1635–1672)*. Leiden: Brill, 2016.

Birkhead, Tim et al, "Willughby's Ornithology," in *Virtuoso by Nature: The scientific worlds of Francis Willughby FRS (1635–1672)*. Edited by Tim Birkhead, 268–304. Leiden: Brill, 2016.

Bloom, Terrie F. "Borrowed Perceptions: Harriot's Maps of the Moon." *Journal of the History of Astronomy* 9 (1978): 117–22.

Bohlin, Diane DeGrazia. *Prints and Related Drawings by the Carracci Family: A Catalogue Raisonné*. Washington, D.C.: National Gallery of Art, 1979.

Bourdieu, Pierre. *Outline of a Theory of Social Practice*. Translated by Richard Nice. Cambridge: Cambridge University Press, 1977.

Broman, Thomas. "Criticism and the Circulation of News: The Scholarly Press in the late Seventeenth Century." *History of Science* 51 (2013): 1–26.

Brown, Harcourt. "History and the Learned Journal." *Journal of the History of Ideas* 33, no. 3 (1972): 365–378.

——. *Scientific Organizations in Seventeenth Century France*. New York: Russell & Russell, 1967.

Bubenik, Andrea. *Reframing Albrecht Dürer: The Appropriation of Art, 1528–1700*. Surrey, UK: Ashgate, 2013.

Bury, Michael. *The Print in Italy, 1550*–1620. London: British Museum, 2001.

Campbell, Mary Baine. *Wonder & Science: Imagining Worlds in Early Modern Europe*. Ithaca: Cornell University Press, 1999.

Castellani, Carlo. "Aldrovandi, Ulisse." In *Dictionary of Scientific Biography*, edited by Charles Gillispie. New York: Scribner, 1970.

Chambers, Douglas D. C. "Evelyn, John (1620–1706)." *Oxford Dictionary of National Biography* (Oxford: Oxford University Press, Sept 2004).

Chartier, Roger, ed. *The Culture of Print: Power and the Uses of Print in Early Modern Europe*. Princeton, N.J.: Princeton University Press, 1989.

Conforti, Maria. "The Experimenters' Anatomy." In *The Accademia del Cimento and its European Context*. Edited by Marco Beretta, Antonio Clericuzio, and Lawrence M. Principe, 31–44. Sagamore Beach: Science History Publications/USA: 2009.

Cook, Harold J. *Matters of Exchange: Commerce, Medicine, and Science in the Dutch Golden Age*. New Haven and London: Yale University Press, 2007.

Cooper, Michael. "Hooke's Career," in *London's Leonardo: The Life and Work of Robert Hooke*, 1–62. Oxford: Oxford University Press, 2003.

Cooper, Thompson. "Collins, Samuel (bap. 1618, d. 1710)." In *Oxford Dictionary of National Biography*. Oxford: Oxford University Press, 2004.

Croft-Murray, Edward, and Paul Hulton. *Catalogue of British Drawings*, Vol. 1, *XVI & XVII Centuries*. London: Trustees of the British Museum, 1960.

Dackerman, Susan. *Prints and the Pursuit of Knowledge in Early Modern Europe*. Cambridge: Harvard Art Museum; New Haven: Yale University Press, 2011.

Daston, Lorraine. "The Empire of Observation, 1600–1800," in *Histories of Scientific Observation*. Edited by Lorraine Daston and Elizabeth Lunbeck. Chicago: University of Chicago Press, 2011, 81–113.

—. "Epistemic Images." In *Vision and Its Instruments* edited by Alina Payne, 13–35. University Park: Pennsylvania State University Press, 2015.

Daston, Lorraine, and Peter Galison. *Objectivity*. New York: Zone Books, 2007.

Daston, Lorraine, and Katharine Park, eds. *Cambridge History of Science*. Vol. 3. *Early Modern Science*. Cambridge: Cambridge University Press, 2006.

Deleuze, Gilles, and Felix Guattari. *On the Line*. Translated by John Johnston. New York: Semiotexte, 1983.

Dennis, Michael Aaron. "Graphic Understanding: Instruments and Interpretation in Robert Hooke's Micrographia." *Science in Context* 3, no. 2 (1989): 309–64.

Dethloff, Diana. "Lely, Sir Peter (1618–1680)." In *Oxford Dictionary of National Biography*. Oxford: Oxford University Press, 2004.

Doherty, Meghan C. "Creating Standards of Accuracy: Faithorne's *The Art of Graveing* and the Royal Society," in *Science in Print: Essays on the History of Science and the Culture of Print*. Edited by Rima D. Apple, Gregory J. Downey, and Stephen L. Vaughn. Madison: University of Wisconsin Press, 2012, 15–36.

—. "Discovering the 'True Form:' Hooke's Micrographia and the Visual Vocabulary of Engraved Portraits." *Notes and Records of the Royal Society of London* 66, no. 2 (2012): 211–234.

—. "Giving Light to Narrative: The Use of Images in Early Modern Journals." *Nuncius: Journal of the Material and Visual History of Science* 30, no. 3 (2015): 543–569.

—. "'Ordinary skill in Cutts:' Visual Translation in Early Modern Learned Journals," in *Translation and the Circulation of Knowledge in Early Modern Science*, edited by Niall Hodson, Sietske Fransen, and Karl Enenkel. Leiden: Brill, 2017.
—. "*The Young-Mans Time Well Spent*: Learning to Draw from a Master," in *Publishing the Fine and Applied Arts, 1500–2000*, edited by Michael Harris, Giles Mandelbrot, and Robin Meyers. London: British Library Press, 2012, 51–78.
Duroselle-Melish, Caroline and David A. Lines. "The Library of Ulisse Aldrovandi (†1605): Acquiring and Organizing Books in Sixteenth-Century Bologna." *The Library*, 7th series, vol. 16, no. 2 (June 2015): 133–161.
Eamon, William. *Science and the Secrets of Nature: Books of Secrets in Medieval and Early Modern Culture*. Princeton: Princeton University Press, 1994.
Edgerton, Samuel Y. *The Heritage of Giotto's Geometry: Art and Science on the Eve of the Scientific Revolution*. Ithaca: Cornell University Press, 1991.
Eisenstein, Elizabeth L. *The Printing Press as an Agent of Change: Communications and Cultural Transformations in Early Modern Europe*. Cambridge: Cambridge University Press, 1979.
'Espinasse, Margaret. *Robert Hooke*. London: William Heinemann, 1956.
Fils, Nathan. "Francis Barlow, the King's Birds, and the Ornithology of Francis Willughby and John Ray." *Huntington Library Quarterly*, 78, no. 2 (Summer 2015): 263–300.
Findlen, Paula. *Possessing Nature: Museums, Collecting, and Scientific Culture in Early Modern Italy*. Berkeley: University of California Press, 1994.
Flick, Gert-Rudolf. *Masters & Pupils: The Artistic Succession from Perugino to Manet, 1480–1880*. London: Hogarth Arts; Paul Holberton Publishing, 2008.
Ford, Brian J. *Images of Science: A History of Scientific Illustration*. London: The British Library, 1992.
Fournier, Marian. *The Fabric of Life: Microscopy in the Seventeenth Century*. Baltimore: Johns Hopkins University Press, 1996.
Fowler, Caroline O., *Drawing and the Senses: An Early Modern History*. Turnhout, Belgium:
Harvey Miller Series in Baroque Art/Brepols, 2017.
Fransen, Sietske. "Antoni van Leeuwenhoek, His images and Draughtsmen." *Perspectives on Science* 27, no. 3 (May–June 2019): 485–544.
Fransen, Sietske and Katherine M. Reinhart. "The Practice of Copying in Making Knowledge in Early Modern Europe: An Introduction." *Word & Image* 35, no. 3 (2019): 211–222.
Fransen, Sietske, Katherine M. Reinhart, and Sachiko Kusukawa. "Copying Images in the Archives of the Early Royal Society." *Word & Image* 35, no. 3 (2019): 256–276.
Freedberg, David. *The Eye of the Lynx: Galileo, His Friends, and the Beginnings of Modern Natural History*. Chicago: University of Chicago Press, 2002.

Gaskell, Roger. "Printing House and Engraving Shop: A Mysterious Collaboration." *The Book Collector* 53 (2004): 213–251.

Godfrey, R.T. "William Faithorne: Portrait Engraver 1616–1691." *History Today* 25, no. 3 (1975): 208–13.

Goldstein, Carl. *Teaching Art: Academies and Schools from Vasari to Albers.* Cambridge: Cambridge University Press, 1996.

Goodwin, Gordon. "Clarke, Timothy (*d.* 1672)," rev. Michael Bevan, *Oxford Dictionary of National Biography,* Oxford University Press, 2004; online ed., Jan 2008.

Greengrass, Mark et al. "Science of the Move: Francis Willughby's Expeditions," in *Virtuoso by Nature: The scientific worlds of Francis Willughby FRS (1635–1672).* Edited by Tim Birkhead. Leiden: Brill, 2016, 142–226.

Griffiths, Antony. "Faithorne, William (c. 1620–1691)." In *Oxford Dictionary of National Biography.* Oxford: Oxford University Press, 2004.

—. *The Print Before Photography: An Introduction of European Printmaking.* London: The British Museum, 2016.

—. *The Print in Stuart Britain, 1603–1689.* London: Published for the Trustees of the British Museum, 1998.

Grindle, Nick. "'No Other Sign or Note Than the Very Order:' Francis Willughby, John Ray and the Importance of Collecting Pictures." *Journal of the History of Collections* 17, no. 1 (2005): 15–22.

Grivel, Marianne. *Le Commerce de l'Estampe à Paris au XVIIe Siècle.* Geneva: Librarie Droz, 1986.

Gross, Joseph. "Ball, William (c.1631–1690)." *Oxford Dictionary of National Biography* (Oxford University Press, 2004; online ed. May 2006).

Hall, A. Rupert. "Medicine and the Royal Society." In *Medicine in Seventeenth Century England* edited by Allen G. DeBus. Berkeley: University of California Press, 1974, 421–452.

Hall, Marie Boas. *Promoting Experimental Learning: Experiment and the Royal Society 1660–1727.* Cambridge: Cambridge University Press, 1991.

Hankins, Thomas L., and Robert J. Silverman. *Instruments and the Imagination.* Princeton, N.J.: Princeton University Press, 1995.

Hanson, Craig Ashley. *The English Virtuoso: Art, Medicine, and Antiquarianism in the Age of Empiricism.* Chicago: University of Chicago Press, 2009.

Hanß, Stefan. "The Fetish of Accuracy: Perspectives on Early Modern Time(s)." *Past and Present* no 243 (May 2019): 267–283.

Harwood, John T. "Rhetoric and Graphics in Micrographia." In *Robert Hooke: New Studies,* edited by Michael Cyril William Hunter and Simon Schaffer, 119–47. Woodbridge, Suffolk: The Boydell Press, 1989.

—. "Science Writing and Writing Science: Boyle and Rhetorical Theory." In *Robert Boyle Reconsidered*, edited by Michael Cyril William Hunter. Cambridge: Cambridge University Press, 1994.

Henderson, Felicity. "Robert Hooke and the Visual World of the Early Royal Society." *Perspectives on Science* 27, no. 3 (May–June 2019): 395–434.

—. "Unpublished Material from the Memorandum Book of Robert Hooke, Guildhall Library MS 1758." *Notes and Records of the Royal Society* 61 (2007): 129–175.

Houghton, Walter E., Jr. "The History of Trades: Its Relation to Seventeenth-Century Thought: As Seen in Bacon, Petty, Evelyn, and Boyle." *Journal of the History of Ideas* 2, no. 1 (January 1941): 33–60.

Hults, Linda C. *The Print in the Western World: An Introductory History.* Madison: University of Wisconsin Press, 1996.

Hume, Julian P. "The History of the Dodo *Raphus Cucullatus* and the Penguin of Mauritius." *Historical Biology* 18, no. 2 (2006): 65–89.

Hunter, Matthew C. "Hooke's Figurations: A Figural Drawing Attributed to Robert Hooke." *Notes and Records of the Royal Society* 64, no. 3 (2010): 251–260.

—. "The Theory of the Impression According to Hooke." in *Printed Images in Early Modern Britain: Essays in Interpretation.* Edited by Michael Hunter. Ashgate: Farnham, Surrey, 2010, 167–190.

—. *Wicked Intelligence: Visual Art and the Science of Experiment in Restoration London.* Chicago: University of Chicago Press, 2013.

Hunter, Michael Cyril William. *Establishing the New Science: The Experience of the Early Royal Society.* Woodbridge, Suffolk: Boydell Press, 1989.

—. *The Image of Restoration Science: The Frontispiece of Thomas Sprat's* History of the Royal Society *(1667).* London: Routledge, 2017.

—. "John Evelyn in the 1650s: A Virtuoso in Quest of a Role." In *Science and the Shape of Orthodoxy: Intellectual Chance in Seventeenth-Century Britain*, 67–98. Woodbridge: The Boydell Press, 1995.

—. *Robert Boyle, 1627–91: Scrupulosity and Science.* Woodbridge, Suffolk: Boydell Press, 2000.

—. *Robert Boyle Reconsidered.* Cambridge: Cambridge University Press, 1994.

Iliffe, Rob. "Material Doubts: Hooke, Artisan Culture and the Exchange of Information in 1670s London." *The British Journal for the History of Science* 28, no. 3 (1995): 285–318.

Ivins, Jr., William M. *Prints and Visual Communication.* Cambridge, MA: The MIT Press, 1953.

Jardine, Lisa. *The Curious Life of Robert Hooke: The Man Who Measured London.* London: HarperCollins, 2003.

Johns, Adrian. "Miscellaneous Methods: Authors, Societies and Journals in Early Modern England." *The British Journal for the History of Science* 33, no. 2 (2000): 159–86.

—. *The Nature of the Book: Print and Knowledge in the Making*. Chicago: University of Chicago Press, 1998.

Jones, Caroline A., and Peter Galison, eds. *Picturing Science, Producing Art*. New York: Routledge, 1998.

Kronick, David A. *A History of Scientific & Technical Periodicals: The Origins and Development of the Scientific and Technical Press, 1665–1790*, 2nd ed. Metuchen, N.J.: 1976.

—. "Notes on the Printing History of the Early *Philosophical Transactions*." *Libraries and Culture* 25, no. 2 (1990): 243–268.

Kusukawa, Sachiko. "*Ad vivum* Images and Knowledge of Nature in Early Modern Europe." In *Ad vivum? Visual Materials and the Vocabulary of Life-Likeness in Europe before 1800* edited by Thomas Balfe, Joanna Woodall and Claus Zittel, 89–121. Leiden: Brill, 2019.

—. "Drawings of Fossils by Robert Hooke and Richard Waller." *Notes and Records of the Royal Society* 67, no. 2 (2013): 123–138.

—. "The Early Royal Society and Visual Culture." *Perspectives on Science* 27, no. 3 (May–June 2019): 350–394.

—. "The *Historia Piscium* (1686)." *Notes and Records of the Royal Society* 54, no. 2 (2000): 179–97.

—. *Picturing the book of nature: image, text, and argument in sixteenth-century human anatomy and medical botany*. Chicago: University of Chicago Press, 2012.

Latham, Robert. "Forward by the General Editor." In *Catalogue of the Pepys Library at Magdalene College Cambridge*, compiled by Eric Chamberlain. Vol. 3, "Prints and Drawings," Part 2, "Portraits." Woodbridge, UK: D.S. Brewer, 1994.

Lefévre, Wolfgang, ed. *Inside the Camera Obscura: Optics and Art under the Spell of the Projected Image*. Preprint 333. Berlin: Max Planck Institute for the History of Science, 2007.

Lefèvre, Wolfgang et al., eds. *The Power of Images in Early Modern Science*. Basel: Birkhäuser Verlag, 2003.

Lloyd, G.E.R. "Saving the Appearances." *The Classical Quarterly* 28, no. 1 (1978): 202–222.

Lincoln, Evelyn. *The Invention of the Italian Renaissance Printmaker*. New Haven: Yale University Press, 2000.

Lind, L. R. *Aldrovandi on Chickens: The Ornithology of Ulisse Aldrovandi (1600) Volume II, Book XIV*. Norman: University of Oklahoma Press, 1963.

Lownes, Albert E. "A Collection of Seventeenth-Century Drawings." *The Auk* 57, no. 4 (1940): 532–535.

——. *Renaissance Books of Science: From the Collection of Albert E. Lownes*, edited by David P. Godine and Owen Gingrich. Hanover: Dartmouth College, 1970.

MacGregor, Arthur. "The Cabinet of Curiosities in Seventeenth-Century Britain." In *The Origins of Museums: The Cabinet of Curiosities in Sixteenth- and Seventeenth-Century Europe*, edited by Oliver Impey and Arthur MacGregor, 147–158. Oxford: Clarendon, 1985.

MacGregor, William B. "The Authority of Prints: An Early Modern Perspective." *Art History* 22, no. 3 (1999): 389–420.

MacLeod, Catharine, and Julia Marciari Alexander. *Painted Ladies: Women at the Court of Charles II*. London: National Portrait Gallery, 2001.

Mandelbrote, Scott. "Ray, John (1627–1705)." In *Oxford Dictionary of National Biography*. Oxford: Oxford University Press, 2005.

Marr, Alexander. "Knowing Images." *Renaissance Quarterly* 69 (2016): 1000–1013.

Marr, Alexander, et al. *Logodaedalus: Word Histories of Ingenuity in Early Modern Europe*. Pittsburgh: University of Pittsburgh Press, 2018.

Massey, Lyle. *Picturing Space, Displacing Bodies: Anamorphosis in Early Modern Theories of Perspective*. University Park: Pennsylvania State University Press, 2007.

McCutcheon, Roger Philip. "The *Journal Des Scavans* and the *Philosophical Transactions of the Royal Society*." *Studies in Philology* 21, no. 4 (1924): 626–28.

Mende, Matthias. *Albrecht Dürer: Zum Leben, Zum Holzschnittwerk, Chronologisches Verzeichnis der Holzschnitte, Hinweise zum aktuellen Schrifttum, Zeittafel, Kondordanz*. Munich: Edition Tomas, 1976.

Millar, Oliver and Diana Dethloff. *Grove Dictionary of Art Online*, s.v. "Lely, Sir Peter." Oxford: Oxford Art Online, 2003.

Moxham, Noah. "Authors, Editors and Newsmongers: Form and Genre in the *Philosophical Transactions* under Henry Oldenburg." In *News Networks in Early Modern Europe*, edited by Noah Moxham and Joad Raymond, 465–494. Leiden: Brill, 2016.

——. "Fit for Print: Developing an Institutional Model of Scientific Periodical Publishing in England, 1665–ca.1714." *Notes and Records of the Royal Society* 69 (2015): 241–260.

Neri, Janice. "Between Observation and Image: Representations of Insects in Robert Hooke's *Micrographia*." In *The Art of Natural History: Illustrated Treatises and Botanical Paintings, 1400–1850*, edited by Amy R. W. Meyers and Therese O'Malley, 83–107. Washington: National Gallery of Art, 2008.

——. *The Insect and the Image: Visualizing Nature in Early Modern Europe, 1500–1700*. Minneapolis: University of Minnesota Press, 2011.

——. "Some Early Drawings by Robert Hooke." *Archives of Natural History* 32, no. 1 (2005): 41–47.

Ochs, Kathleen H. "The Royal Society of London's History of Trades Programme: An Early Episode in Applied Science." *Notes and Records of the Royal Society of London* 39, no. 2 (April 1985): 129–158.

Olson, Roberta J. M. and Alexandra Mazzitelli. "The Discovery of a Cache of over 200 Sixteenth-Century Avian Watercolors: A Missing Chapter in the History of Ornithological Illustration." *Master Drawings*, 45, no. 4 (Winter, 2007): 435–521.

Oxford English Dictionary, s.v. "accurate," "accuracy," "accurately," and "accurateness." OED Third Edition, December 2011.

Pace, Claire. *Grove Art Online*, s.v. "Disegno e colore." Oxford: Oxford Art Online, 2003. https://doi-org.berea.idm.oclc.org/10.1093/gao/9781884446054.article.T022879. Accessed March 18, 2019.

Parish, Joylon C. *The Dodo and the Solitaire: A Natural History*. Bloomington: Indiana University Press, 2013.

Park, Katharine. *Secrets of Women: Gender, Generation, and the Origins of Human Dissection*. New York: Zone Books, 2006.

Parshall, Peter. "Graphic Knowledge: Albrecht Dürer and the Imagination." *Art Bulletin* 95, no. 3 (2013): 393–410.

—. "*Imago Contrafacta*: Images and Facts in the Northern Renaissance." *Art History* 16, no. 4 (1993): 554–579.

Payne, Alina, ed. *Vision and Its Instruments*. University Park: Pennsylvania State University Press, 2015.

Phillips, John C. "Leonard Baldner, Seventeenth Century Sportsman and Naturalist: An Unrecorded Copy of His Book, Containing His Portrait." *The Auk* 42, no. 3 (1925): 332–341.

Pointon, Marcia. *Hanging the Head: Portraiture and Social Formation in Eighteenth-Century England*. New Haven: Yale University Press, 1993.

Poole, William. "The Willughby Library in the Time of Francis the Naturalist." In *Virtuoso by Nature: The scientific worlds of Francis Willughby FRS (1635–1672)*, edited by Tim Birkhead, 227–243. Leiden: Brill, 2016.

Poole, William, Felicity Henderson, and Yelda Nasifoglu. "Robert Hooke's Books." Accessed October 29, 2020. https://hookesbooks.yeldanasifoglu.com/.

Poovey, Mary. *A History of the Modern Fact: Problems of Knowledge in the Sciences of Wealth and Society*. Chicago: University of Chicago Press, 1998.

Purver, Margery. *The Royal Society: Concept and Creation*. Cambridge: M.I.T. Press, 1967.

Raven, Charles E. *John Ray, Naturalist, His Life and Works*. Cambridge: The Univerity Press, 1942.

Reinhart, Katherine M. "Richard Waller and the Fusion of Visual and Scientific Practice in the Early Royal Society." *Perspectives on Science* 27, no. 3 (May–June 2019): 435–484.

Richardson, Ruth. *Death, Dissection, and the Destitute.* 2d ed. Chicago: University of Chicago Press, 2000.

Roberts, Lissa, Simon Schaffer, and Peter Dear, eds. *The Mindful Hand: Inquiry and Invention from the Late Renaissance to Early Industrialisation.* Amsterdam: KNAW, 2007.

Rostenberg, Leona. *The Library of Robert Hooke: The Scientific Book Trade of Restoration England.* Santa Monica, CA: Modoc Press, Inc., 1989.

Ruestow, Edward G. *The Microscope in the Dutch Republic: The Shaping of Discovery.* Cambridge: Cambridge University Press, 1996.

Sargent, Rose-Mary. *The Diffident Naturalist: Robert Boyle and the Philosophy of Experiment.* Chicago: University of Chicago Press, 1995.

Scala, Gail Ewald. "An Index of the Proper Names in Thomas Birch 'A History of the Royal Society' (London, 1756–1757)." *Notes and Records of the Royal Society* 28, no. 2 (1974): 263–329.

Shapin, Steven. *A Social History of Truth: Civility and Science in Seventeenth-Century England.* Chicago: University of Chicago Press, 1994.

Shapin, Steven, and Simon Schaffer. *Leviathan and the Air-Pump: Hobbes, Boyle, and the Experimental Life.* Princeton, N.J.: Princeton University Press, 1985.

Shapiro, Barbara J. *A Culture of Fact: England, 1550–1720.* Ithaca, N.Y.: Cornell University Press, 2000.

Shuffelton, George. *Codex Ashmole 61: A Compilation of Popular Middle English Verse.* The Camelot Project at the University of Rochester. Kalamazoo, Mich.: Medieval Institute Publications, 2008.

Sloan, Kim. *'A Noble Art': Amateur Artists and Drawing Masters, c.1600–1800.* London: British Museum Press, 2000.

—. "Sir Hans Sloane's Pictures: The Science of Connoisseurship or the Art of Collecting?" *Huntington Library Quarterly* 78, no. 2 (2015): 381–415.

Smith, Pamela H. "Art, Science, and Visual Culture in Early Modern Europe," *Isis* 97 (2006): 83–100.

—. *The Body of the Artisan: Art and Experience in the Scientific Revolution.* Chicago: University of Chicago Press, 2004.

Stinjman, Ad. *Engraving and Etching, 1400–2000: A History of the Development of Manual Intaglio Printmaking Processes.* Leiden: Brill, 2012.

Strauss, Walter L. *Sixteenth Century German Artists: Albrecht Dürer.* Vol. 10. *The Illustrated Bartsch.* New York: Abaris Books, 1980.

Stresemann, Erwin. *Ornithology from Aristotle to the Present.* Translated by Hans J. and Cathleen Epstein. Cambridge: Harvard University Press, 1975.

Swan, Claudia. "*Ad Vivum, Naer Het Leven,* from the Life: Defining a Mode of Representation." *Word and Image* 11, no. 4 (1995): 353–372.

Taylor, E.G.R. "Robert Hooke and the Cartographical Projects of the Late Seventeenth Century (1666–1696)." *The Geographical Journal* 90, no. 6 (1937): 529–540.

Thackray, John C. *A Catalogue of Portraits, Paintings, and Sculpture at the Natural History Museum, London*. London: Mansell, 1995.

Turner, Gerard L'E. "The Impact of Hooke's *Micrographia* and its influence on microscopy." In *Robert Hooke and the English Renaissance*, edited by Paul Kent and Allan Chapman, 124–145. Herefordshire: Gracewing, 2005.

Van Helden, Albert. "'Annulo Cingitur': The Solution of the Problem of Saturn." *Journal for the History of Astronomy*, 5 (1974): 155–74.

—. "Christopher Wren's '*De Corpore Saturni*'." *Notes and Records of the Royal Society of London*, 23, no. 2 (Dec., 1968): 213–229.

—. "Saturn and his Anses." *Journal of the History of Astronomy* 5 (1974): 105–121.

Walpole, Horace. *A Catalogue of Engravers, Who Have Been Born, or Resided in England*. 2d ed. London: Printed for J. Dodsley, 1786.

Wilson, Catherine. *The Invisible World: Early Modern Philosophy and the Invention of the Microscope*. Princeton, N.J.: Princeton University Press, 1995.

Winkler, Mary G., and Albert van Helden. "Representing the Heavens: Galileo and Visual Astronomy." *Isis* 83, no. 2 (1992): 195–217.

—. "Johannes Hevelius and the Visual Language of Astronomy." In *Renaissance and Revolution: Humanists, Scholars, Craftsmen, and Natural Philosophers in Early Modern Europe*, edited by Judith Veronica Field and Frank A. J. L. James, 95–114. Cambridge: Cambridge University Press, 1993.

Wragge-Morley, Alexander. *Aesthetic Science: Representing Nature in the Royal Society of London, 1650–1720*. Chicago: University of Chicago Press, 2020.

Yale, Elizabeth. *Sociable Knowledge: Natural Knowledge and the Nation in Early Modern Britain*. Philadelphia: University of Pennsylvania Press, 2016.

Index

References to illustrations are in *italics*

Accademia dei Lincei, *Melissographia*, bees 37–39, *39*
accuracy
- as additive process 180
- best practices 83
- components 35–36, 215
- and consensus 35, 216
- as cultural construct 32
- and engraver's burin 30, 31, 219
- and Fellows of the Royal Society 49, 89
- illusion of 26, 79, 197
- intaglio prints 36
- and microscopic world 98
- and neatness 79
- and *Philosophical Transactions* 180
- protocols 216
- regimes 216
- and scientific development 32, 40
- and truth 118
- *see also* engraving accuracy

accurate
- definition 31
- etymology 32
- occurrences 32

"ad vivum" term, use 34
air compression research, van Leeuwenhoek 207–8
air pump, (Boyle) *28*, 81, *82*, 83, *84*, 85–86, 219
Aldrovandi, Ulisse 141
- collections 150–51
- *Ornithologiae* 150–51
  - bird images 165, *166*, *168*, *169*
  - Ray's borrowings 151, 165

Alpers, Svetlana 114
anatomy
- articles in *Philosophical Transactions* 201–9
- image representation 181

Andrea del Sarto 77
art, and science, closeness 219
art books 52–53
- audience for 54–55

artisans, collaboration with Fellows of the Royal Society 26–27
artists' manuals 12
astronomy, image representation 181
Aubry, Claude, broadside 204–5, *204*

Bacon, Francis 53
- *Advancement of Learning* 13

Baldner (Baltner), Leonard 159–60, 160–61
Ball, William
- *Saturn* *185*
- observations 186–88
- *Sketch of Saturn* 189, *189*

Bate, John, *The Mysteryes of Nature and Art* 54, 72
- title page *73*

Beale, John 30
Bennett, Jim 105
Birch,, *History of the Royal Society* 113, 117, 201
Birkhead, Tim 142
black ibis *167*, 169–71
blood transfusion experiments, Clarke *202*, 203
A Book of Drawing: A Young-Mans Time Well Spent 53, 54, 56
- drawing tools 66
- Dürer connection 62–63
- editions 62
- *How to draw faces* *91*
- human body 67
- images to copy 66, *67*
- observations of masters' works 68
- patterns 67
- purpose 64–66
- sources 63

Borel, Pierre, *Observationum Microcospicarum Centuria*, butterfly's antennae 108, *109*
Bosse, Abraham 52, 71
- meeting with Faithorne 78
- *Traicté des manières de graveur en taille douce* 77

Bourdieu, Pierre, on habitus 26
Boyle, Robert
- *A continuation of new experiments physico-mechanical* *28*
- air pump *28*, 81, *82*, 83, *84*, 85–86
- commitment to experimentation 80–81
- *Medicina Hydrostatica*, experiments 56, 81
- *New Experiments Physico-Mechanicall*, experimental set-ups 56
- portrait (Faithorne) 18, *19*, 81, *82*

Browne, Alexander 63
- *Ars Pictoria* 115

Buonanni, Filippo, *Micrographia Curiosa*, *A Flea* *129*
burin *see* engraver's burin
butterfly, antennae 108, *109*

camera obscura, Hooke's interest in 114
Campani, Giuseppe, observation of Saturn 193–94
Carracci, Agostino 63
Casseri, Giulio, *Pentaestheseion, hoc est De qvinqve sensibvs liber* *25*

Cassini, Jean-Dominique, observations on
    Saturn 194–95, *195*, 196–97, *198*
Cesi, Federico 39, 107
Charles II, King, portrait (Faithorne) *119*
    armor/collar reflection 120, 127
    hair representation 118–19
Clarke, Timothy 199, 201
    blood transfusion experiments *202*, 203
Clusius, Carolus, *Exoticorum libri decem* 146
"collections of anomalies" term 35
Collins, Samuel, *A Systeme of Anatomy* 15, *17*,
    20–21
    plates 21
    "Preface to the Tables" 21, 22
communications networks
    Royal Society of London 219
    Saturn research 189–90, 190–91, 194, 199
consensus, and accuracy 35
"counterfacta" term 34
Cousin, Jean 63
    Livre de Pourtraittre 52
Cowley, Abraham 36
    ode "To the Royal Society" 33, 67–68

Dackerman, Susan 12, 29
Daston, Lorraine, on epistemic image 29
Daston, Lorraine & Galison, Peter 35
de bondt, Jacob, *De medicina Indorum* 146
de Graaf, Regnier 199, 201
    research on testicles 205–6, *208*
Delaram, Francis, *Portrait of Albrecht Dürer* *64*
Deleuze, Gilles & Guattari, Felix 215
dodo
    description 145–46
    images 146, *147–48*, *150*
    sources on 146–50
drawing (limning)
    A Book of Drawing (Jenner) 115
    Ars Pictoria (Browne) 115
    books on 115
    Carracci brothers' prints 52
    Diagraphia, sive Ars Delineatoria (Jans-
        sonius) 52
    Evelyn on 62
    Hooke, on learning 55, 115
    How to draw faces *65*
    and knowledge production 62
    learning 55
    Livre de Pourtraittre (Cousin) 52
    manuals 54, 57, 63, 64
    prestige 52
    and printmaking 65
    Underweysung der Messung (Dürer) 52
    and visual culture 215
Dürer, Albrecht
    drawing prodigy 57
    Evelyn on 60
    model for learning to draw 63
    portrait (Delaram) *64*
    Underweysung der Messung 52

Eamon, William 72
Edwards, George 173
engraver's burin 11, 14, 80, 86, *87*, 216
    and accuracy 30, 31, 219
    handle 86, *87*, 88
    and image representation 180–81, 197,
        199
    key instrument 218
    whetting 82–83
    see also graver
engraving
    craft knowledge 52
    etching, combination 30–31, 31fn63
    history, in *Sculptura* (Evelyn) 77
    manuals 52
    "tyranny of the rule" 27
    visual vocabulary 101
    see also intaglio prints
engraving accuracy 27
    and the burin 30, 31
    Hooke 40, 99, 130
    Ornithology (Willughby) 140
    Ray 141–42, 144, 169
    visual effects of 40
Ent, George 98
etching, engraving, combination 30–31,
    31fn63
Evelyn, John
    on Boyle 217
    as connoisseur 69–71
    on drawing 62
    Grand Tour in Italy 71
    portrait (Nanteuil) *70*
    print collection 76
        Sculptura 53, 61
        dedication to Evelyn 217
        on Durer 60
        engraving history 77
        exemplary works 76
        Frontispiece *75*
        on Hevelius 50–51, 71
        How to draw a sphere and a head *59*
        on origins of sculpture 74–75
        purpose 69–70, 217–18
        on study of good models 76–77
    The Hatcher *61*
    on three-dimensionality 57–58, 59–60
    translation, *Idée de la Perfection de la*
        Peinture 13
    View of Naples from Mount Vesuvius *72*
    on Villamena 76

"fact" term, origin 34
Faithorne, William
    A Human Body Opened *17*
        compared with Villiers portrait 22–23
    book and print seller 115
    meeting with Bosse 78
    Portrait of Barbara Villiers 15, *16*, 18
    Portrait of Charles II 118–20, *119*

Portrait of Robert Boyle 18, *19*, 81, *82*
Pepys on 19–20
The Art of Graveing and Etching 53, 54, 56
accuracy 79
best practices 79–80
error recognition 79
judging one's work 80
plates 86
purpose 78–79
varnish use 85
Walpole on 77–78
Fellows of the Royal Society 13
collaboration with artisans 26–27
and image accuracy 49, 89
and image production 36
reading of art books 53
see also Royal Society of London
Fialetti, Odoardo 63
flea *129*
fly, *Eye of a Fly* image *111*, 112
Ford, Brian 102
Fowler, Caroline 54
Fréart, Roland, *Idée de la Perfection de la Peinture* 13

Galileo 183
Gessner, Konrad 141
authority on bird names 153
Historia animalium 150, 171, *172*
golden eagle
description 153, 155, 157
images *154–56*
graver
handle 86, *87*, 88
meaning 79
use 86, *87*
see also engraver's burin
great grey gull 162, *163–64*, 165
Grew, Nehemiah, on Willughby's *Ornithology* 173
Griendel, Johann, *Micrographia* 130
Mold Spores *131*
Griffiths, Anthony 29, 30

habitus
Bourdieu on 26
Hooke 113
Hall, A. Rupert 200
Hanß, Stefan 31
Hanson, Craig 76
Haydocke, Richard, translation, Lomazzo's *Trattato dell'arte della pittura* 30
Hevelius, Johannes
Evelyn on 50–51, 71, 196
Lunar Eclipse observed 20 November 1668 *50*
Saturn, observations 195–96, *196*
Selenographia 51
Hilliard, Nicholas, *Treatise Concerning the Arte of Limning* 54
Hodierna, Giovanni Battista
Eye of a Fly, image *111*, 112
L'Occhio della Mosca 111, 125
Hollar, Wenceslas 116
Hooke, Robert 80
accuracy 40, 99, 130
arbiter of Royal Society drawings 115
book collection 116
camera obscura, interest in 114
copyists 128, 130
Curator of Experiments (Royal Society) 99, 114
Drawing of Experimental Setup 127, *129*
Felt-makers at work *60*
habitus 113
insect drawings 126–27, *128*
on learning to draw 55, 115
and London arts community 105
Micrographia 12, 41
deer hair image 99, *100*, 105
experimental set-ups 56
fly head image 36–37, *38*
Pepys on 127
preface 41
scale images *110*
sensory evidence 102
shadowing advice 116
size ratios 111
spider image 103, *104*
working method 98
part transcription of *Traicté des manières de graveur en taille douce* (Bosse) 115
print buying 116
seeing differently 98, 98–99, 99, 101
Three Heads and Two Figure Studies 89, *90*
on viewing Mars 182
visual vocabulary 118
Hume, Julian 146
Hunter, Matthew 115
Wicked Intelligence *105*
Hunter, Michael 69
Huygens, Christiaan
observations on Saturn 191, *192*, 193
Saturn's ring hypothesis 181, 183–84, 193, 198–99
Systema Saturnium 199

image production 11
image representation
anatomy 181
astronomy 181
and the burin 180–81, 197, 199
sexual reproduction 181
images
accuracy 11, 49
consumption of 54, 55
epistemic, definition 29
functionality 29
"life-like" 140
mimetic funtions 14

production, and Fellows of the Royal Society 36
as scientific *lingua franca* 217
sources 12
value of 14, 216
see also insect images
insect images
Hooke 126–27, *128*
microscopy *108*, *109*
intaglio printing 77
intaglio prints 11, 12, 30, 34, 142
accuracy 36
creating 79, 89
Ivins, William 27

Janssonius, Joannes 63
Diagraphia, sive Ars Delineatoria 52
Jenner, Thomas, *A Book of Drawing* 115
Journal des Sçavans 178, 180
Joyliffe, George 205

King, Edmund 199, 206, 206–7
knowledge production
and drawing 62
and *Philosophical Transactions* 179–80
Kusukawa, Sachiko 26, 34, 140

Lely, Peter, Court Painter to Charles II 113–14
limning *see* drawing
Lister, Martin 141, 142
London, print market 29–30

MacGregor, William 55
Martyn, John, Royal Society Printer 143
Michelangelo 76
microscope 219
and accuracy 98
Pepys' use 102
seeing differently 97–98, 98, 98–99, 99, 101
microscopy
butterfly's antennae *109*
Eye of a Fly *111*
interpretation issues 102, 117–18
scale images *110*
three-dimensionality 123, 130
visual culture 106–12
visual vocabulary 112–13
Weevil *108*
Moray, Robert, Sir 189, 190
Moxham, Noah 180

"naer het leven" term, use 34
Nanteuil, Robert, *Portrait of John Evelyn* *70*
networks *see* communications networks
Nieremberg, Juan Eusebio, *Exoticorum libri decem* 146

observation 33
Ogilby, John 116
Oldenburg, Henry 51, 144, 177–78, 178–79, 181

Parshall, Peter 34
Payne, Alina 31
Peacham, Henry, *The Art of Drawing with the Pen* 66
Pepys, Samuel
on Fairthorne's portrait of Barbara Villiers 19–20
on *Micrographia* (Hooke) 127
use of microscope 102
perspective, and depiction of objects 59–60
Philosophical Transactions 12
and accuracy 180
anatomy articles 201–9
beginning 18
format 180
illustrations 42
and knowledge production 179–80
policy 178–79
purpose 177–78
Saturn observations *186*, 191, 194, *195*
visual epistemologies 180
Piso, Willem, *De Indiae utriusque re naturali et medica* 146
Poovey, Mary 34
poppy seeds *122*, *124*
Hooke/Power images compared 123, 125–26
portraiture
idealized versus veristic image 103
visual vocabulary 103, 126–27, *128*
Power, Henry, *Experimental Philosophy* 121–22
Corn Poppy Seeds *122*
Ribban *121*
print, and modern science 40
print market, London 29–30
printmaking, and drawing 65
prints
role in visual education 55
see also intaglio prints
publishing, scientific 180

Raven, Charles 142
Ray, John 41
bird history, correction methods 144, 166, 169
bird identification 144–45
dodo description 145–46
engraving accuracy 141–42, 144, 169
golden eagle, description 153
Historia Plantarum 140
methodology 141, 162, 164, 166–67
preface to *Ornithology* (Willughby) 137–38, 141
print images, reliance on 157
visual judgment 144
Willughby's executor 142
on Willughby's sources 137–38
see also Willughby, Francis
Robert, Lissa, *The Mindful Hand* 15

Royal Society of London
  Baconian project 32, 151, 179, 200
  dissections 22
  foundation 11
  histories of 13, 33, 113, 117, 201
  history of trades project 13, 53, 56, 69
  see also Fellows of the Royal Society

Sargent, Rose-Mary 81–82
Saturn
  Balls' observations 186–88, 189–90
  Balls' sketch 188, 189–90, *189*
  Cassini's observations 193–94, 194–95, *195*, 196–97, *198*
  communications network on 189–90, 190–91, 194, 199
  "figure", articles about 184, 191
  Galileo's observations 183
  Hevelius' observations 195–96, *196*
  Huygens' observations 191, *192*, 193
  Huygen's ring hypothesis 181, 183, 193, 198–99
  observations in *Philosophical Transactions* *186*, 191, 194, *195*
  problems of viewing 182–83
  visualizations *185*, *186*
  Wren's theory 190
Savery, Roelandt, *Dodo* *150*
"scheme", meaning 21
Schongauer, Martin 76
science
  and art, closeness 219
  development
    and accuracy 32, 40
    and print 40
sculpture, origins, Evelyn on 74–75
semen, van Leeuwenhoek's research 208–9, *209*
sexual reproduction, image representation 181
Shapin, Steven 31
Shapin, Steven & Schaffer, Simon 21–22, 27
  virtual witnessing 187
Shapiro, Barbara 34
Sloan, Kim 52
smew (white nun)
  images 157, *158–59*, *161*
  species duplication 160, 161
Smith, Pamela 12, 15
Sprat, Thomas, *History of the Royal Society* 13, 33
standard
  definition 31
  see also truth
Stelluti, Francesco
  Melissographia *39*
  Weevil image *108*
Swan, Claudia 34

testicles, research 205–6, *208*
three-dimensionality 101, 127
  Evelyn on 57–58, 59–60
  microscopy 123, 130
Tradescant's Cabinet 146
truth
  and accuracy 118
  definition 31–32
  see also standard
"truth to nature" term 35
Tyson, Edward 22

Urban III, Pope 37

van Calcar, Jan Stephan, *Anatomy of Female Torso* *24*
van Helden, Albert 183, 196
van Horne, Johannes 201
van Leeuwenhoek, Antoni 27, 182, 199
  air compression research 207–8
  semen research 208–9, *209*
Vasari, Giorgio
  on Florentine *disegno* 30
  Lives of the Painters 76
Vesalius, Andreas, *De Humani Corporia Fabrica* 23–24, *24*
Villamena, Francesco, Evelyn on 76
Villiers, Barbara, portrait (Faithorne) 15, *16*, 18
virtual witnessing, Shapin & Schaffer 187
visual culture, and drawing 215
visual education, role of prints 55
visual vocabulary
  engraving 101
  Hooke's 118
  microscopy 112
  portraiture 103, 126–27, *128*

Walpole, Horace
  A Catalogue of Engravers 77
  on Faithorne 77–78
Wharton, Thomas 205
Wilkins, John 98
Willughby, Francis
  commonplace book 69
  Ornithology 12, 41, 218
    accuracy 140
    Aldrovandi, borrowings from 151, 165
    bird classification 143
    engravers 143
    golden eagles 153, *154–56*, 155, 157
    great grey gull 162, *163–64*, 165
    origins 142
    plates *139*, *152*
    smew (white nun) 157, *158–59*, 160, 161–62, *161*
    source materials 171
  see also Ray, John
Wilson, Catherine 106
Wren, Christopher, Sir, Saturn theory 190

For Product Safety Concerns and Information please contact our EU
representative GPSR@taylorandfrancis.com
Taylor & Francis Verlag GmbH, Kaufingerstraße 24, 80331 München, Germany

www.ingramcontent.com/pod-product-compliance
Ingram Content Group UK Ltd.
Pitfield, Milton Keynes, MK11 3LW, UK
UKHW021722240926
475702UK00029BA/1760
* 9 7 8 1 0 4 1 1 7 8 7 8 1 *